PROGRESS IN BIOORGANIC CHEMISTRY

PROGRESS IN BIOORGANIC CHEMISTRY

VOLUME TWO

Edited by

E. T. KAISER

Departments of Chemistry and Biochemistry
University of Chicago

F. J. KÉZDY

Department of Biochemistry
University of Chicago

A WILEY-INTERSCIENCE PUBLICATION

JOHN WILEY & SONS

NEW YORK · LONDON · SYDNEY · TORONTO

Library of Congress Cataloging in Publication Data:

Main entry under title:

Progress in bioorganic chemistry.

(Progress in biorganic chemistry series)
Includes bibliographies.
1. Biological chemistry—Collected works. 2. Chemistry, Physical organic—Collected works. I. Kaiser, Emil Thomas, 1938–   ed. II. Kézdy, F. J., 1929–   ed.

QD415.AlP76   547′.1′3   75–142715
ISBN 0–471–45486–9 (V. 2)

Printed in the United States of America

10 9 8 7 6 5 4 3 2 1

CONTRIBUTORS

STEPHEN J. BENKOVIC, *Department of Chemistry, The Pennsylvania State University, University Park, Pennsylvania*

WILSON P. BULLARD, *Department of Chemistry, The Pennsylvania State University, University Park, Pennsylvania*

GERALD CLEMENT, *Department of Biological Chemistry, Hahnemann Medical School, Philadelphia, Pennsylvania*

E. H. CORDES, *Department of Chemistry, Indiana University, Bloomington, Indiana*

CARLOS GITLER, *Centro de Investigacion y de Estudios Avanzados, Mexico, D.F.*

MICHAEL H. KLAPPER, *Department of Chemistry, Division of Biochemistry, The Ohio State University, Columbus, Ohio*

FROM THE
PREFACE TO THE SERIES

Bioorganic chemistry is a new discipline emerging from the interaction of biochemistry and physical organic chemistry. Its origins can be traced to the enzymologists whose curiosity was not satisfied with the purification and the superficial characterization of an enzyme, to the physical organic chemists who had the conviction that the elementary steps of biological reactions are identical to those observed in organic chemistry, and to the physical and organic chemists who wished to understand and to imitate *in vitro* the unequaled catalytic power and specificity exhibited by living organisms. As with all interdisciplinary sciences, bioorganic chemistry uses many of the methods and techniques of the disciplines from which it is derived; many of its protagonists qualify themselves as physical organic chemists, enzymologists, biochemists, or kineticists. It is, however, a new science of its own by the criterion of having developed its own goals, concepts, and methods.

The principal goal of bioorganic chemistry can be defined as the understanding of biological reactions at the level of organic reaction mechanisms, that is, the identification of the basic parameters which govern these reactions, the formulation of quantitative theories describing them, and the elucidation of the relationships between the reactivity and the structures of the molecules participating in the process. This definition is narrower than one which some scientists would give. They might prefer to include areas such as medicinal chemistry, for example, as part of the field of bioorganic chemistry. Accordingly, the goals which they would cite would differ from those which we have considered. We do not seek here to argue or to defend our concept of what constitutes bioorganic chemistry, but within the framework of our definition we believe that there is a real distinction between much work in present day medicinal chemistry and that in the bioorganic field. In our conception of bioorganic chemistry the emphasis is on mechanism.

The theoretical formulation of the understanding of biochemical and, therefore, enzyme-catalyzed reactions has required the elaboration of new concepts, such as multifunctional catalysis, stereospecificity by three-point attachment, and control of reactivity by conformational changes. Many of the new concepts will not survive; they will be redefined, discarded, or reevaluated as fortunately always happens in science. But the trend is clearly apparent—these new concepts are providing us with efficient tools of great power which can be used to describe and discuss enzymatic reactions.

As to the methods of bioorganic chemistry, they are conceptually the same as those for the study of any chemical reaction; they include analytical and physical techniques. However, the complexity of the reacting molecules has resulted in methods which are new and unique in their ability to probe the chemistry of a functional group surrounded by a multitude of very similar groups or a chemical event accompanied by a host of satellite reactions. The discovery of numerous methods involving active-site directed reagents, "reporter molecules," and chromophoric substrates illustrates the usefulness and the elegance of the new science.

The future of bioorganic chemistry appears very promising, and the fields to cover in the future are immense and unexplored. The earliest work has concentrated on the understanding of general acid–general base catalyzed reactions, hydrolytic reactions, and the role of proteins in enzymatic catalysis. The mechanism of enzymatic catalysis by most coenzymes is very far from being well described, and the very prominent role of metal ions in catalysis is only beginning to emerge. Other important problems, such as surface catalysis at biological membranes, transport mechanisms, the process by which ribosome-catalyzed reactions occur, and the reactivity of RNA and DNA molecules, are at an early stage of development or are completely unexplored.

As a result, a rapid growth of bioorganic chemistry is desirable and is currently underway, as evidenced by the large number of papers published on the subject. An unfortunate result of this rapid expansion is the scarcity of comprehensive treatments of bioorganic chemistry. The rapid progress in this field makes it likely that large portions of any comprehensive textbook will become obsolete soon, although the student of bioorganic chemistry may still learn some of the basic concepts of the subject from them. Because of many factors, it would be possible to revise textbooks only at infrequent intervals. For this reason the format of presenting comprehensive treatments of limited subjects seemed more appropriate to us. It would provide the investigators, interested readers, and students with a thorough and critical evaluation of those aspects of bioorganic chemistry where definite and substantial progress has been achieved.

It is the hope of the Editors of this series to be able to respond to the need for up-to-date comprehensive treatments of important topics in bioorganic chemistry. In attempting to do so we would like to provide treatments of bioorganic subjects which will be general enough to retain the attention of most workers in the field and which will be at a level beyond that of a usual review article or literature survey. Since many aspects of bioorganic chemistry are still in the process of evolution, we also would like to provide a forum where the authors can express challenging new ideas and present stimulating and, frequently, controversial discussions. For this reason we hope to give the authors somewhat more latitude than is customary in this kind of publication, while still retaining the requirement of scientific sobriety.

E. T. KAISER
F. J. KÉZDY

Chicago, Illinois
January 1971

PREFACE TO VOLUME TWO

In the present volume of *Progress in Bioorganic Chemistry* two chapters are devoted to physical-chemical aspects of biologically important systems. In the first one, the relation between micellar structure and reactivity is reviewed, whereas the second one is concerned with the thermodynamics of peptide-peptide and protein-solvent interactions. Recent progress and theories on enzymatic and model reactions involving folic acid are discussed in the third chapter. The final chapter is devoted to the study of reactions catalyzed by pepsin, a model enzyme for biological hydrolytic reactions in strongly acid media.

By this choice of topics, we wish to emphasize the breadth of the field that is encompassed by bioorganic chemistry.

We are grateful to the authors who contributed to this present volume and to Mrs. Hanna Posner for her help.

Chicago, Illinois   E. T. KAISER
October 1972   F. J. KÉZDY

CONTENTS

REACTION KINETICS IN THE PRESENCE OF MICELLE-FORMING SURFACTANTS

E. H. CORDES

Indiana University, Bloomington, Indiana

CARLOS GITLER

Centro de Investigacion y de Estudios Avanzados, Mexico, D.F.

1

1 INTRODUCTION

In 1936, G. S. Hartley introduced the term *amphipathy* to describe the unusual properties of aqueous solutions of detergent molecules [1]:

This unsymmetrical duality of affinity is so fundamental a property of paraffin-chain ions, being directly responsible for all the major peculiarities of paraffin-chain salts in aqueous solutions, that it is worth while, if only for emphasis, to give it a special name. The property is essentially the simultaneous presence (in the same molecule) of separately satisfiable *sym*pathy and *anti*pathy for water. I propose, therefore, to call this property amphipathy—the possession of both feelings.

There are many molecules in biological systems that possess this property. Thus nearly all components of cellular membranes, such as phospholipids, glycolipids, cholesterol, and long-chain fatty acid salts, are clearly amphipathic molecules. In addition, polypeptide chains can be included in this category because they contain within the same molecule hydrophobic and hydrophilic side chain residues.

In aqueous solutions these molecules exhibit their amphipathic nature in several ways. First, they can adsorb at the air-water or oil-water interface so that the apolar portions occupy the air or oil phase while the polar groups are retained in the aqueous environment. Second, the molecules can associate to form micellar or liquid crystalline aggregates in which the apolar molecules are directed away from water in close van der Waals contact, while the polar groups are directed so that they have maximum contact with water. Third, in the case of long polypeptide chains, the molecules can assume a three-dimensional conformation such that the majority of the apolar side chains are removed from the aqueous environment, while the polar side chains are oriented to maximize their interactions with water.

In all three cases, the segregation of hydrophobic and hydrophilic groups leads to the formation of interfaces between essentially apolar regions and water. Following Hartley, we refer to these as *amphipathic surfaces or interfaces*. It is clear from the preceding discussion that the surface regions of enzymes and membranes contain amphipathic interfaces. The purpose of this chapter is to draw attention to the unique properties of such interfaces and to attempt to relate them to specific functions.

A great many biological processes occur on or involve amphipathic surfaces. These include enzyme-catalyzed reactions, assembly of multienzyme complexes, facilitated transport of ions and molecules, cellular adhesion and recognition, generation and conduction of nervous impulses, and sensory reception and transduction, to name a few. Understanding of these phenomena is of great general importance for several reasons, not the least of which is the fact that associated malfunctions result in numerous pathological states. In part, development of this understanding at a molecular basis necessarily involves much surface chemistry, a field that has experienced a marked deemphasis in the last two decades.

Initial steps directed toward sorting out those factors that are responsible for the characteristics of chemical processes at pertinent surfaces include additional understanding of the structure of the surfaces themselves and of the behavior of simple organic molecules adsorbed thereto. The necessary structural studies are being actively pursued on a number of fronts, and we refer to a number of important results. But our attention focuses on the kinetics of organic reactions occurring at just one such surface, that of micelles formed from relatively simple surfactants. One cannot hope to gain great insight into the physiological processes noted above through the study of such simple models, although we draw repeated parallels between these systems and enzyme-catalyzed reactions. Nevertheless, valuable information concerning them may result from extension of this work to more complicated systems that more nearly mimic the physiological ones. For the moment, reaction kinetics at micellar surfaces is an "interface" between physical organic chemistry on the one hand and biochemistry and physiology on the other. In the last section of this chapter, efforts to develop the nature of this interface are made. For the moment, let us examine some typical amphipathic molecules and the structures derived from them as a prelude to more detailed considerations of micelles themselves and of the properties of reactions occurring in their presence.

2 AMPHIPATHIC MOLECULES AND ASSOCIATED STRUCTURES

There are a number of structurally distinct classes of amphipathic molecules, and these form a variety of kinds of structures. The simplest amphipathic molecules are constituted from a straight chain of carbon atoms, usually 8–18 in number, to which is attached a polar group that may be anionic, cationic, zwitterionic, or nonionic. Typical surfactants in this class include:

$$CH_3(CH_2)_n—OSO_3^-$$ Anionic

$$CH_3(CH_2)_n—\overset{+}{N}(CH_3)_3$$ Cationic

$$CH_3(CH_2)_n—\overset{+}{N}(CH_3)_2—CH_2CH_2CH_2—OSO_3^-$$ Zwitterionic

$$CH_3(CH_2)_n—\langle\ \rangle—O—(CH_2CH_2O)_m—H$$ Nonionic

Of course, great structural variation is possible even within this general class of molecules because both the length of the hydrophobic portion and the nature of the hydrophilic head group, as well as its position along the backbone, may be varied. The properties of the aggregates formed from these surfactants and the conditions under which they are formed depend, of course, on all these parameters.

As the concentrations of surfactants in an aqueous solution are increased, many of the physical and chemical properties of the solution change rather

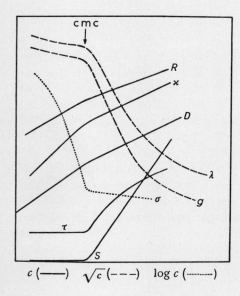

$c\ (\text{———})$ $\sqrt{c}\ (---)$ $\log c\ (\cdots\cdots\cdots)$

Figure 1 Typical variations in the physical properties of an ionic association colloid. Characteristic plots shown are R, refractive index; D, density; χ, specific conductance; τ, turbidity; S, solubility of a water-insoluble dye such as Orange OT (all against the concentration c of the association colloid); g, osmotic coefficient; λ, equivalent conductance (against $\sqrt{c}$); σ, surface tension (against $\log c$) [P. Mukerjee, *Advan. Colloid Interface Sci.*, **1**, 241 (1967)].

abruptly (but continuously) over a characteristic narrow concentration range, termed the *critical micelle concentration* (cmc). Affected properties include the refractive index, conductivity, viscosity, surface tension, turbidity, capacity to solubilize hydrocarbons and dyes, and the like; the way that some of these quantities change with changing surfactant concentration is shown in Figure 1. Over the narrow concentration range centering about the cmc, individual surfactant molecules form aggregates, usually containing 20–100 individual molecules [2]:

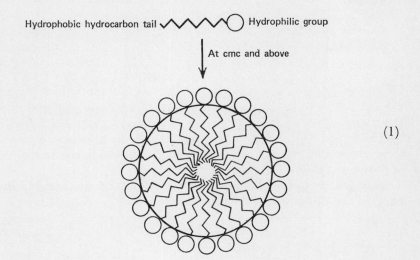

(1)

As indicated in equation 1, these aggregates are usually spherical (exceptions will be dealt with subsequently); they are organized so that the hydrophobic chains are hidden within the structure and thus isolated from the aqueous environment and the hydrophilic head groups are exposed on the micellar surface. Values for the cmc and for the aggregation number for typical surfactants and the derived micelles are collected in Table 1.

The thermodynamics of micelle formation has been considered on the basis of a number of models including mass action, phase separation, and spherical double-layer theory. A number of excellent discussions of these theories are available [2–8]. Suffice it to say that the principal driving force for micelle formation appears to be entropic, not enthalpic, deriving from the negative free energy change accompanying the liberation of water molecules in the immediate vicinity of exposed hydrophobic chains when micelles are formed [10]. The same driving force is important in determining the three-dimensional structure of protein molecules [11].

Simple surfactants are capable of forming larger and more complicated structures as well. For example, hexadecyltrimethylammonium bromide in water at 27° exists as spherical micelles at a surfactant concentration of 0.05 M but as large rod-shaped micelles at twice this concentration [12]. Such micelles may have molecular weights of a million or more. At still higher concentrations, a liquid crystalline phase is formed from infinitely long rod-shaped micelles that occupy positions at the corners of a regular hexagonal lattice [12]. Similar transitions in micellar structure may be induced at low surfactant concentrations by employing high concentrations of certain counterions [13, 14].

Micelles are not static species but are in dynamic equilibrium with co-existing monomers. A number of independent measurements indicate that the rate constants for dissociation of individual surfactant molecules from the micelle are large and may be in the $10^3 \, sec^{-1}$ range [15–17].

Because micelles formed from simple surfactants are our particular concern in the kinetic experiments, further consideration of their properties is deferred at this point; structural details and the nature of interactions with organic molecules are provided subsequently.

TABLE 1 PARAMETERS FOR MICELLE FORMATION FROM SIMPLE IONIC SURFACTANTS[a]

Surfactant	Added Salt	cmc $\times 10^3$ M	Aggregation Number
Sodium dodecyl sulfate	0.050 M NaCl	2.30	84
	0.201 M NaCl	0.94	107
	0.506 M NaCl	0.51	126
Sodium hexadecyl sulfate	—	0.52	—
Dodecyltrimethylammonium bromide	0.050 M NaBr	5.71	72
	0.100 M NaBr	3.88	73
	0.508 M NaBr	1.46	88
Dodecyltrimethylammonium chloride	0.050 M NaCl	9.50	57
	0.201 M NaCl	4.36	62
	0.506 M NaCl	2.48	68
Dodecyltrimethylammonium nitrate	0.050 M NaNO$_3$	4.70	63
	0.101 M NaNO$_3$	3.32	73
	0.253 M NaNO$_3$	2.04	80
	0.509 M NaNO$_3$	1.40	86
Tetradecyltrimethylammonium bromide	—	3.5	75
Hexadecyltrimethylammonium bromide	—	0.92	61

[a] The bulk of the data are from Emerson and Holtzer [7].

A more complicated example of a surfactant is provided by the lecithins:

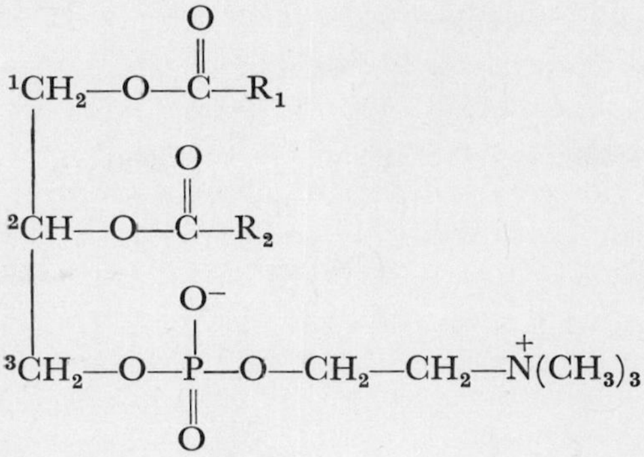

$$
\begin{array}{l}
\overset{\displaystyle O}{\overset{\displaystyle \|}{{}^1CH_2-O-C-R_1}} \\[2mm]
\overset{\displaystyle O}{\overset{\displaystyle \|}{{}^2CH-O-C-R_2}} \\[2mm]
{}^3CH_2-O-\underset{\underset{\displaystyle O}{\displaystyle \|}}{\overset{\overset{\displaystyle O^-}{\displaystyle |}}{P}}-O-CH_2-CH_2-\overset{+}{N}(CH_3)_3
\end{array}
$$

In lecithins isolated from natural sources, the fatty acid esterified at the 1 position is usually saturated and contains 16 or 18 carbon atoms; that esterified at the 2 position is usually unsaturated, but has about the same chain length. A number of synthetic lecithins have been prepared that are homogeneous with respect to fatty acid content. The structural chemistry of lecithins dispersed in water is richer than that for the simpler surfactants. Preparations that are swollen in water, mechanically dispersed in water, or sonically dispersed in water form an array of structurally distinct phases [18–22]. Certain of these are diagrammed in Figure 2. The simplest of these is the monolayer formed at the air-water interface in which the polar head is immersed in the water and the hydrophobic tail sticks up in the air. The remaining structures have in common a bimolecular leaflet arrangement, commonly thought to be characteristic of biological membranes (see, however, [23]), in which the polar groups are outward directed and the hydrophobic ones are inward directed. A theoretical analysis of the stability of various phospholipid modifications has been provided [24].

A distinct type of biologically important surface-active molecule is provided by the bile salts; three typical examples are shown on page 9. Note that the polar and nonpolar portions of these molecules are not isolated from one another as they are in the surfactants previously considered. One of the polar groups does occupy a terminal position on the side chain of the steroid nucleus. Either two or three hydroxyl groups, however, are also present in these structures, and these are all disposed at one side of the molecule. Hence one face of the steroid nucleus has pronounced polar character, and the other, nonpolar character. As a consequence, micelles formed from these surfactants tend to be small, containing 2–4 molecules

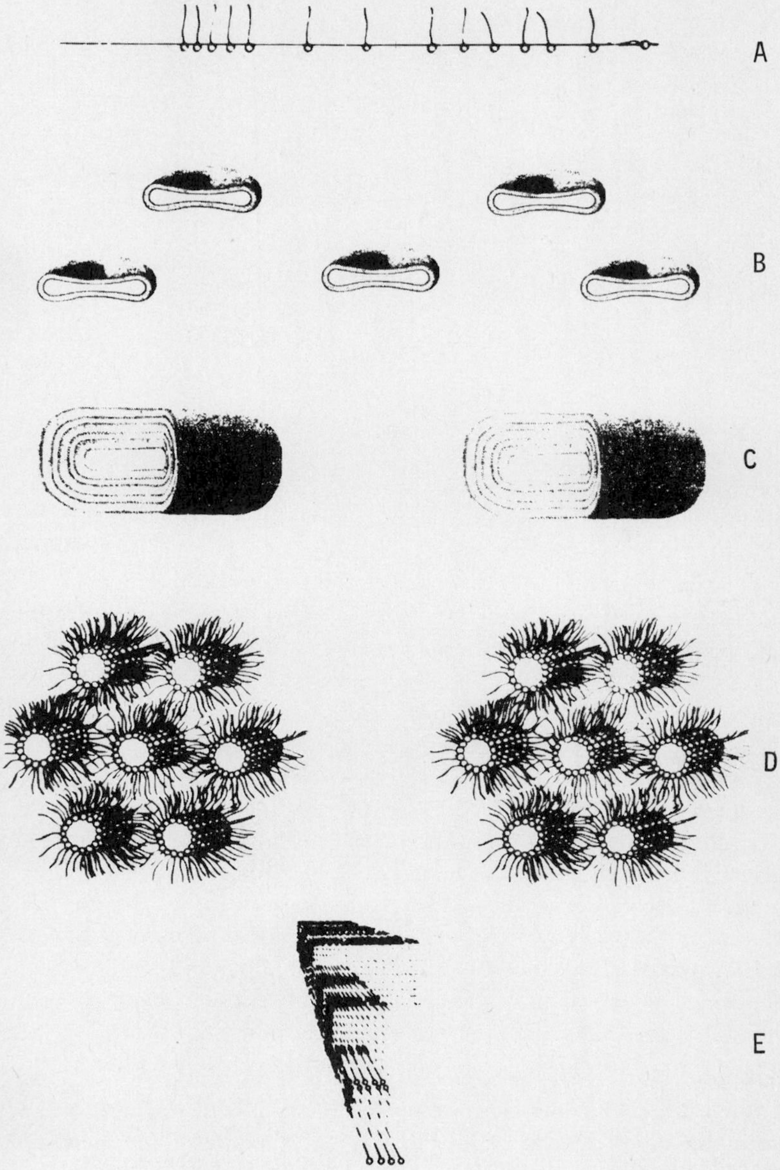

Figure 2   Diagrammatic representation of molecular arrangements of phospholipids. (A) Monolayer model-air/water interface. (B) Sonicated smectic mesophase. (C) Smectic mesophase model. (D) Hexagonal mesophase-limited water. (E) Anhydrous liquid crystal. (Not to any scale) [A. D. Bangham, *Progr. Biophys. Mol. Biol.*, **18,** 29 (1968)].

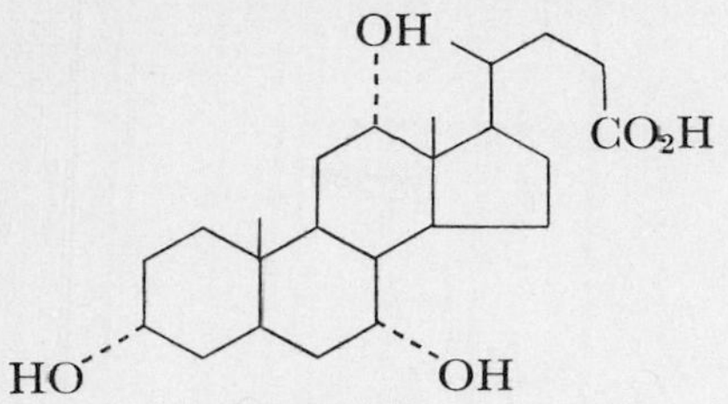

Cholic acid

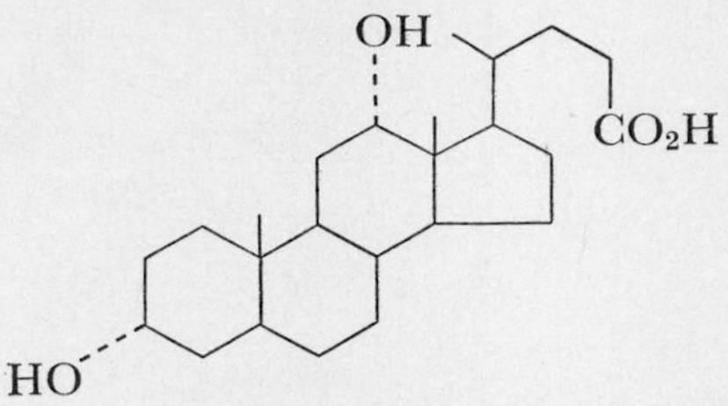

Deoxycholic acid

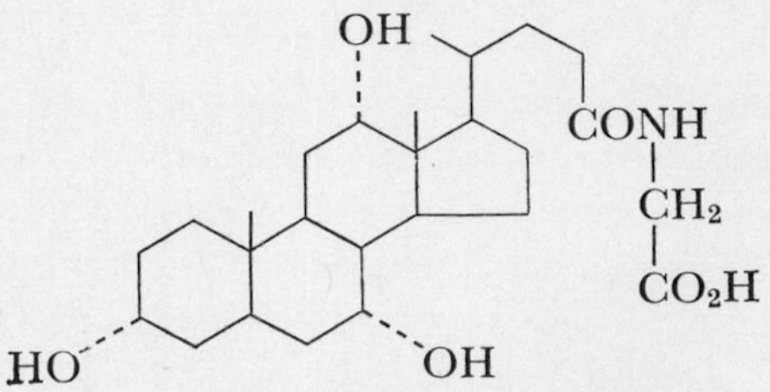

Glycocholic acid

per micelle, and have their nonpolar faces pressed together, isolated from the solvent [25–27]:

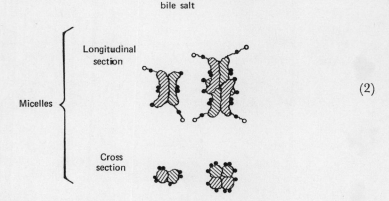

$$\tag{2}$$

Certain bile salts in the presence of high concentrations of salts do form very much larger micelles that are probably assembled through interactions between the exposed polar groups as the water available for their solvation is decreased [27].

Polysoaps provide a type of structure related to those discussed previously but in which the properties of surfactants are merged with those of polyelectrolytes. Typical polysoaps are synthesized by quaternization of polyvinylpyridine with, for example, dodecyl bromide and ethyl bromide [28].

The resulting structure:

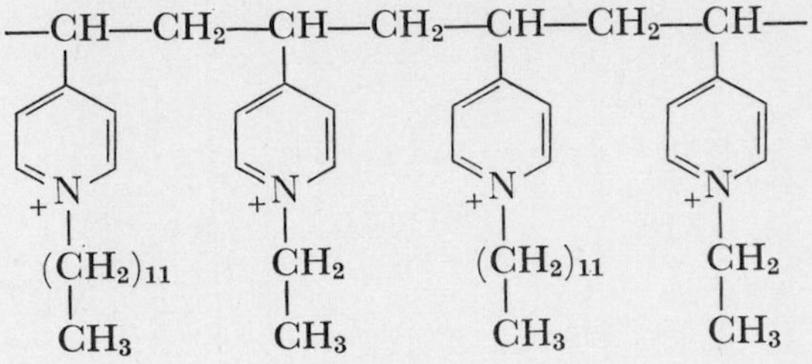

$$—CH—CH_2—CH—CH_2—CH—CH_2—CH—$$

$+N$	$+N$	$+N$	$+N$
$(CH_2)_{11}$	CH_2	$(CH_2)_{11}$	CH_2
CH_3	CH_3	CH_3	CH_3

may be thought of as a series of simple surfactant molecules held together through their attachment to a polymethylene chain. These molecules, under the proper conditions, fold up in such a way as to maximize the number of charged groups exposed to solvent and to minimize the number of hydrophobic groups so exposed. Thus their organization is basically related to that of a simple micelle, and the same driving forces appear to account for the formation of compact structures in the two cases.

Polysoaps provide a structural link between simple surfactant molecules and other polymers, including proteins. Like polysoaps, proteins consist of a backbone this time polar in character, from which are dangled side chains ranging in their characteristics from charged to polar uncharged to apolar. The results of x-ray diffraction studies of protein structure establish that in this case, too, the molecular organization is such that the polar groups occupy positions on the molecular exterior and the hydrophobic ones occupy the interior, insofar as is possible [29].

3 THE AMPHIPATHIC SURFACE

One way to demonstrate the presence of amphipathic interfaces and to monitor their behavior is the study of the spectroscopic changes of dyes adsorbed into proteins, membranes, and surfactant micelles. Studies of this type have been reviewed by Horton and Koshland [30] and by Edelman and McClure [31]. Recently, attempts have been made to determine quantitatively the binding site polarity using as probes molecules that fluoresce intensely when bound to amphipathic surfaces or when dissolved in nonpolar solvents, but which have extremely low fluorescence in aqueous solutions. Turner and Brand [32] have correlated the fluorescence yield, emission maximum and band width of N-arylaminonaphthalene sulfonates in a variety of solvents with Z values that represent an empirical solvent polarity scale suggested by Kosower [33]. It is concluded by these authors that the reciprocal of the wavelength of maximum fluorescence $(\bar{\nu}_F)$ of

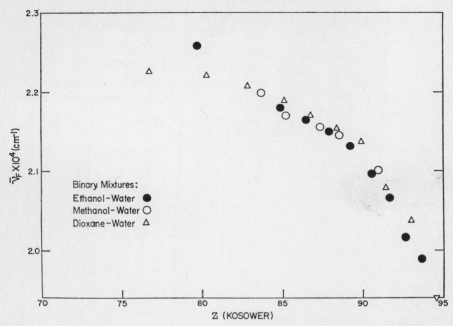

Figure 3 Frequencies of the fluorescent emission maximum of ANS plotted against the Kosower Z values for three mixed solvent systems [D. C. Turner and L. Brand, *Biochemistry*, **7,** 3381 (1968)].

the adsorbed probes gives the most reliable estimates of binding site polarity. Figure 3 shows the values of the reciprocal of the emission maximum, $\bar{\nu}_F$, for 1-aminonaphthalene-7-sulfonate (1,7-ANS) dissolved in three organic solvent-water binary mixtures plotted as a function of solvent Z value. It can be observed that a relatively smooth curve results. If the dye is allowed to bind to different amphipathic interfaces, the measured values of $\bar{\nu}_F$ can be used to estimate the Z value of the environment surrounding the bound dye. These authors further show that the $\bar{\nu}_F$ values are invariant over the pH range of 1.5–12.0 in D_2O as a function of solvent viscosity. Table 2 shows the estimated polarities for 1,7-ANS bound to different proteins, erythrocyte membranes, and micelles of cationic and nonionic surfactants. The average Z values lie between 80 and 88 for almost all proteins studied and for the erythrocyte membranes and the surfactant micelles. Since 1,7-ANS is an amphipathic molecule, it is likely that it is located in an apolar-polar interface, and the values obtained give an indication that the polarity of these sites is definitely lower than that of water ($Z = 94.6$) and approaches that of pure ethanol ($Z = 79.6$). The similarity in the values for hexa-decyltrimethylammonium bromide and for Triton X-100 indicate that the

TABLE 2   ESTIMATION OF THE POLARITY OF BINDING SITES FROM THE EMISSION MAXIMUM OF BOUND 1,7-ANS[a]

1,7-ANS Bound to	Protein Concentration (mg/ml)	$\bar{\nu}_F \times 10^4$ (cm^{-1})	Z (Estimated)
Alcohol dehydrogenase (horse liver)	2.0	2.145	89
Alcohol dehydrogenase (yeast)	0.8	2.169	86.5
Glutamate dehydrogenase	1.20	2.193	84
Lactate dehydrogenase	1.5	2.179	85.5
Chymotrypsinogen	0.9	2.198	84
Chymotrypsin	1.1	2.100	91
Trypsin	5	2.193	84
Phosphofructokinase	1.5	2.193	84
Hexokinase	0.19	2.193	84
Luciferase	0.1	2.242	74–81
Myokinase	0.5	2.227	77–81.5
Ribonuclease	2.2	2.252	73–80
Deoxyribonuclease	0.76	2.193	84
Adenosine deaminase	0.2	2.227	81
Enolase	0.75	2.212	82.5
Aldolase	3.0	2.212	82.5
Phosphoglucose isomerase	0.40	2.208	82.5
Lysozyme	1.0	2.155	88
Plasma albumin (bovine)	3.0	2.208	82.5
Hemoglobin-free erythrocyte membranes	1.0–2.0	2.214	82.5
Hexadecyltrimethylammonium bromide (15 mM)	—	2.174	85.5
Tetradecyldimethylbenzyl ammonium chloride (50 mM)	—	2.183	84.5
Triton X-100 (5 mM)	—	2.183	84.5

[a] Data on proteins from Turner and Brand [32]. Data on membranes and surfactants from Gitler [34].

probe is not sensitive to the surface charge. However, no interaction was observed between 1,7-ANS and anionic surfactants, indicating that charge is important for binding to the micelle. Turner and Brand have shown further that similar values of Z can be derived for the protein binding sites using as probes 1-anilinonaphthalene-5-sulfonate and 2-toluidinonaphthalene-6-sulfonate. On the other hand, 1-anilinonaphthalene-8-sulfonate gave lower Z values for the proteins and for the membranes and surfactant micelles, indicating that specific interactions might also be present.

In this context it is of interest to mention that the dye ethidium bromide

(3,8-diamino-6-phenyl-5-ethylphenanthridinium bromide) binds with enhanced fluorescence to double-stranded DNA and RNA [35], to erythrocyte and mitochondrial membranes, and to micelles of sodium lauryl sulfate [36]. On the other hand, binding with low enhancement in the fluorescence occurs with single-stranded RNA, denatured DNA, or polyvinylsulfate [35]. Also, no fluorescence is observed with sodium lauryl sulfate at detergent concentrations below the critical micelle concentration. Ethidium is a molecule in which the positive charge is delocalized throughout the conjugated double bonds of the aromatic rings. This results in the dye being soluble in solvents such as ethylene glycol, ethanol, butanol, and octanol, where it shows enhanced fluorescence over that in water. From the foregoing it is concluded that the increase in quantum yield on binding is the result of the immersion of the dye in a hydrophobic region of the native nucleic acid, of the membranes, and of the surfactant micelles where it is protected against quenching by the aqueous solvent.

These observations clearly indicate the presence of amphipathic interfaces in a series of biological macromolecules or macromolecular aggregates. They serve to draw attention to one of the unique properties of these interfaces, mainly the presence of apolar regions in equilibrium with the aqueous solvent. In the following pages we attempt to describe other characteristics of amphipathic interfaces, using mainly as a model the studies performed with surfactant micelles.

The amphipathic character of the surfaces of micelles, membranes, and proteins is, of course, manifested in many ways other than that reflected in the fluorescence behavior of adsorbed dyes. We examine some of these in our subsequent discussion of micellar organization. But we may note in passing that amphipathic characteristics are reflected rather directly in terms of interaction of these surfaces with ions on the one hand and with nonpolar organic molecules on the other. Enzymes provide a particularly good example because they are known to interact strongly with (and their biological activity is influenced strongly by) both ions and hydrocarbons. It is particularly striking that surfactants, which combine the two characteristics, interact with proteins so strongly that they frequently cause denaturation of the protein at concentrations in the range $0.001-0.01\ M$; the best simple ionic denaturants must ordinarily be present at concentrations greater than $1\ M$ to effect a comparable degree of disruption of the protein structure.

4 MICELLES

Because we are concerned with micelles formed from simple surfactant molecules with respect to reaction kinetics, we now consider the structural organization of such micelles in somewhat more detail; a schematic view

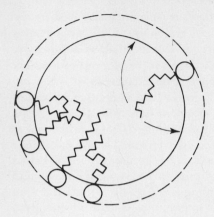

Figure 4   A schematic view of a micelle formed from an ionic surfactant.

of a micelle formed from, for example, sodium dodecyl sulfate or dodecyltrimethylammonium bromide, is provided in Figure 4. The core of the micelle is composed of the hydrophobic chains of the surfactant molecules. Surrounding the core is the *Stern layer*, or *palisade layer*, which is occupied by the regularly spaced charged head groups of the ionic micelle together with a certain number of counterions. The exterior boundary of the Stern layer corresponds closely with the micellar shear surface. Hence the counterions within the Stern layer, which are sufficient to neutralize 50–70 percent of the surface charges [37, 38], are tightly bound and form part of the kinetic micelle. The remainder of the counterions are less tightly associated with the micelle and form a Gouy-Chapman layer several hundred angstroms in thickness.

There are three potentials of note in micelles. The phase boundary potential is the total difference, which may be several hundred millivolts, between the micellar core and the solution. The Gouy-Chapman and zeta potentials are, respectively, those between the exterior surface of the Stern layer and between the shear surface and the bulk solution; these are usually not very different. The potential drop across the micellar surface may be several ten thousand volts per centimeter because the interface extends over only a short distance.

With this brief introduction, let us look somewhat more closely at the properties of typical micelles formed from ionic surfactants.

4.1 The Micellar Core

Most of the available evidence suggests that the micellar core has properties related to those of liquid hydrocarbons. This point of view is supported by observations concerning the thermodynamics of micelle formation, the polarization of fluorescence emission of apolar molecules dissolved in the micellar interior, micellar heat capacities, and the compressibility of micelles.

Table 3 shows the standard free energies of transfer from aqueous to different non polar environments per methylene and benzene group. In the case of the surfactants, the values calculated by Molyneux et al. [39] and Shinoda [46] are based on the free energy contribution to micellization in aqueous solutions of homologous series of surfactants, assuming that the total free energy change is divisible into independent additive contributions from the component parts of the amphiphile molecule. The values for the standard free energy contributions to the binding of peptide inhibitors to the active site of papain have also been shown, to a first approximation, to be additive.

The results shown for the contribution to micellization in aqueous solutions yield closely similar values per methylene group for nine amphipathic molecules varying widely in the nature of the ionic group present. This constancy suggests that the internal structures of the micelles formed by these various amphiphiles are similar. The fact that the value of $\Delta G/$ ($-CH_2-$) for the micellization (0.65 ± 0.02 kcal mole^{-1}) is lower than that for the transfer of an alkane from aqueous solution into the pure liquid alkane (i.e., 0.83 ± 0.02) and is in fact much closer to the value obtained for the adsorption at the air-water interface appears to indicate, as Molyneux

TABLE 3 STANDARD FREE ENERGY CHANGES FOR TRANSFER OF METHYLENE OR BENZENE GROUPS FROM AQUEOUS TO VARIOUS NONPOLAR ENVIRONMENTS

	$-\Delta G$ (kcal/mole)	System	Reference
Per methylene group	0.62–0.65	Adsorption at the air-water interface	40
	0.65 ± 0.02	Contribution to micellization in aqueous solutions	39
	0.73	Partitioning between alcohol and water	41
	0.81–0.85	Partitioning between hydrocarbon and water	42–44
	1.0	Binding of inhibitors to the active site of papain	45
Per benzene ring	2.25	Contribution to micellization in aqueous solutions	46
	1.92	Partitioning between ethanol and water	41
	3.0	Binding of inhibitors to the active site of papain	45

et al. [39] have suggested, that when micellization occurs, although the internal structure of the micelle may approximate that of liquid hydrocarbon, the amount of hydrocarbon surface exposed to the aqueous phase is merely reduced rather than completely eliminated [2].

Recently results have been obtained that indicate that accurate measurements of the viscosity in the hydrocarbon core of the micelles may be obtained from studies of the polarization of the fluorescence emission of apolar molecules dissolved in the micelle interior [47]. For example, 2-methylanthracene (2-MeA), which is nearly insoluble in water, is readily dissolved by micelles of sodium dodecyl sulfate. The fluorescence emission of the 2-MeA in the micellar solution is almost completely depolarized and nearly equal to that of the 2-MeA dissolved in liquid tetradecane. This indicates that there is essentially the same freedom for the rotation of the 2-MeA in the two environments. The absence of polarization of the fluorescence is

TABLE 4 VALUES OF MICROVISCOSITY FOR MICELLES DERIVED FROM SEVERAL SURFACTANTS[a]

Surfactant	Concentration (M)	Probe[b]	p^c ($\times 100$)	$\bar{\eta}\,(27°)^d$ (cP)	E^e (kcal/mole).
Dodecyltrimethylammonium bromide	5×10^{-2}	2MA	5.38	26	7.2
		Per	2.08	17	9.6
Tetradecyltrimethylammonium bromide	2×10^{-2}	2MA	5.10	32	7.1
		Per	2.18	21	9.5
Hexadecyltrimethylammonium bromide	10^{-2}	2MA	5.73	30	6.2
		Per	2.05	19	9.6
Octadecyltrimethylammonium bromide	10^{-2}	2MA	3.77	50	6.1
		Per	3.06	37	8.0
n-Dodecane				1.33	2.8
n-Hexadecane				2.98	3.8
White oil U.S.P. 35				124	12.7

[a] Microviscosities derived from the degree of depolarization and the excited state lifetime of the dye. All data are from Ref. 47.
[b] 2MA is 2-methylanthracene, excited at 382 nm, and Per is perylene, excited at 413 nm.
[c] Degree of polarization at 27°.
[d] Microviscosity; values for dodecane, hexadecane and white oil, included for comparison, are from F. D. Rossini, *Selected Values of Properties of Hydrocarbons*, National Bureau of Standards, pp. 111, 112 (1947).
[e] Fusion activation energy, derived from the change in microviscosity with temperature in the range 4–27°.

significant since the 2-MeA has a relatively short lifetime of the excited state (3.5 ± 0.09 nsec in tetradecane and 3.3 ± 0.1 nsec in the micelle interior). A summary of this study is provided in Table 4. It is of interest that fluorescent molecules—such as sodium 1-anilinonaphthalene-8-sulfonate, sodium N,N-dimethyl-1-aminonaphthalene-5-sulfonate, fluorescein, acridine orange, and rhodamine B that, being charged, would be expected to be localized in the palisade layer of the surfactant micelles—have been found to emit highly polarized fluorescence [48].

Goddard et al. have measured the partial molal heat capacity of the $(CH_2)_6$ group in the micelle of potassium octanoate; a value of 0.55 cal deg^{-1} gm^{-1} was obtained [49]. This is substantially the same as the value of 0.53 cal deg^{-1} gm^{-1} found for several liquid hydrocarbons, supporting the concept of a liquid hydrocarbon-like micellar interior.

Finally, Shigehara has measured the compressibilities of micelles derived from several alkyl sulfates employing ultrasonic interferometry [50]. These values, too, provide support for the thesis under development.

The only serious challenge to this thesis derives from the interesting studies of Muller and his associates, who have prepared several surfactants having terminal trifluoromethyl groups [51–53]:

$$CF_3—(CH_2)_n—CO_2^-Na^+$$
$$CF_3—(CH_2)_n—SO_3^-Na^+$$

These workers have established that these surfactants behave in a manner similar to those possessing a normal terminal methyl group. Studies of the chemical shifts of the fluorine atoms reveal that their environment in the micelles is "wet" [51, 52] suggesting that the micellar interior contains appreciable amounts of water. However, it is also possible that the observations are a consequence of the rather polar character of the trifluoromethyl group. As is developed subsequently, even weakly polar molecules tend to be solubilized on the micellar exterior rather than the interior. Therefore the trifluoromethyl groups of the surfactant molecules tend to occupy positions on the micellar surface, to the extent that this is possible, exposed to the solvent, accounting for the observed chemical shifts. Other explanations are also possible.

4.2 The Micellar Surface

In our discussion of amphipathic surfaces generally, we have pointed out several properties of the micellar surface. The nature of this surface may be brought into sharper focus, however, by the following considerations.

The thickness of the Stern layer is about the same as that of the ionic headgroups that occupy it, and the hydration of these groups is similar to that of the isolated charged groups themselves [54, 55]. The number of counterions that are bound within the Stern layer is a sensitive function of the nature and concentration of those in the environment. Increasing concentration and increasing counterion hydrophobicity both increase the fraction of charges at the micellar surface that are neutralized by inclusion of counterions.

The surface of micelles formed from ionic surfactants is rough [37, 56]. Proton nmr measurements and other evidence indicate that a portion of the hydrocarbon chain of the surfactant is, transiently at least, exposed to the aqueous solvent regardless of the nature of the micellar interior [57]. Indeed, it is precisely this roughness that gives the micellar exterior its amphipathic character, the capacity to interact with both ions and un-chaged hydrophobic molecules.

Finally, our conclusions concerning the polar character of the micellar surface derived from the fluorescence measurements are neatly confirmed by elegant studies of Mukerjee and Ray, who have measured the wavelength maximum of the charge transfer band formed between the headgroups of N-alkylpyridinium ions and several anions [58]. From this information, they conclude that the dielectric constant at the surface of micelles formed from such surfactants is near 36.

4.3   Solubilization of Organic Molecules by Micelles

The first stage involved in catalysis of an organic reaction by micelles is the interaction of the substrate with the micelle. Clearly, the characteristics of the adsorption process will have a major influence on the kinetics of the reaction under consideration. In the first place, the adsorption isotherm should account for the dependence of reaction rate on micelle and substrate concentrations. In addition, the localization of the adsorbed substrate in or on the micelle determines its reactivity by yielding or denying access of other reagents to it, by establishing the electric field and local dielectric constant at the site of reaction, and so forth. We can imagine several ways in which an organic substrate might be associated with a micelle: it might occupy the micellar interior; a nonpolar portion could occupy the interior with a polar moiety exposed on the surface; or the molecule might occupy the exterior surface itself in a variety of ways. A large number of methods have been employed for the measurement of adsorption isotherms and localization of organic molecules in micellar systems. Both the methods and the results derived from their use have been extensively reviewed [59–61]. Typical results, pertinent to the discussion of kinetics that follows, are collected in Table 5. The central point to be derived from data of this type is that molecules of appreciable polarity occupy the micellar surface, not the micellar

TABLE 5 LOCATION OF SOLUBILIZATES IN MICELLAR SYSTEMS[a]

Method	Surfactant	Solubilizate	Suggested Location of Solubilizate	Reference
UV spectroscopy	Potassium dodecanoate Dodecylammonium chloride Polyoxyethylene(23)dodecanol (Brij 35)	Ethylbenzene	Micelle core	59
		Naphthalene Anthracene trans-Azobenzene	Partly in hydrocarbon core and partly near the polar head groups	60, 61
		Dimethyl phthalate	Micelle surface	
PMR spectroscopy	Hexadecyltrimethylammonium bromide	Benzene N,N-Dimethylamine Nitrobenzene	Micelle-water interface	
		Isopropylbenzene	Oriented at the micelle-water interface	
Combined potentiometric, UV, and PMR techniques	Polyoxyethylene (20-24) hexadecanol (Cetomacrogol) Sodium dodecyl sulfate	Benzoic acid	Junction of the hydro-carbon core and the hydrated oxyethylene layer of the micelle	62, 63
Combined ESR, PMR, and UV techniques	Sodium dodecyl sulfate	Stable radicals	Random spatial orientation in a dynamic time-averaged equilibrium	64

[a] Reproduced from Fendler and Fendler [65].

19

interior. Substantially all the organic molecules of interest with respect to the reaction kinetics either are sufficiently polar to occupy the micellar surface or contain at least one functionality that it expected to do so.

5   REACTION KINETICS IN MICELLAR SYSTEMS

Although the chemical literature during the first half of this century includes a number of reports dealing with reaction kinetics in solutions containing ionic and nonionic micelles, Duynstee and Grunwald in 1959 were the first clearly to point out those factors that are likely to underlie the rate effects observed [68]. Since that time, a substantial number of related studies have appeared, and the rate of their appearance is increasing. It has been established that, for a variety of reactions, micelles formed from ionic surfactants induce rate increases, generally between the limits of ten- and one hundredfold. Fendler and Fendler have provided a comprehensive review of effort in this field [65]. Our efforts in this direction are limited to pointing out several of the central features of these reactions, following the format of a short previous review[69].

The kinetics of organic reactions occurring in micellar systems are dominated by two factors: *electrostatic* interactions and *hydrophobic* interactions between the micellar phase and reactants, transition states, and products. Prior to examining the individual aspects of these systems, let us focus briefly on these generalizations.

Kinetic studies performed thus far in solutions of micelle-forming surfactants can be grouped into two classes. The first includes those cases in which the surfactant provides a medium for the reaction but does not participate directly in it. For example, the kinetics of hydrolysis of methyl orthobenzoate are very substantially altered by the addition of small concentrations of sodium dodecyl sulfate to the reaction medium, although this surfactant does not react with the ortho ester. The second class includes those reactions in which the surfactant does participate directly in the reaction, either as a catalyst or as a substrate. Thus the kinetics of acid-catalyzed hydrolysis of sodium dodecyl sulfate are markedly modified upon micellation of this material. In a related case, the nucleophilicity of long-chain N-acyl histidines toward p-nitrophenyl esters is enhanced upon the incorporation of the histidines into micelles. In the former case, the surfactant is itself the substrate and is consumed in the course of the reaction; in the second, the surfactant is transiently modified through interaction with the substrate but is subsequently regenerated. As we shall see, the basic characteristics of both classes of reaction are related. We may also note at this point that, generally, one of the substrates will be uncharged and, hence, will interact with the micellar phase by virtue of its hydrophobic properties, and another will be charged, hence, interact electrostatically with this phase. With these points in mind, then, we turn to the general aspects of these reactions noted above.

The electrostatic basis of kinetic effects in micellar systems may be appreciated on the basis of the following few examples. First, the acid-catalyzed hydrolysis of sodium alkyl sulfates is markedly promoted by micellation of the substrates; the uncatalyzed reaction is unaffected; and the base-catalyzed reaction is strongly inhibited [70–72]. Second, the hydrolysis of the dianions of 2,4- and 2,6-dinitrophenyl phosphates is promoted by cationic surfactants but unaffected by anionic or nonionic surfactants [73]. Third, the acid-catalyzed hydrolysis of methyl orthobenzo-ate is subject to catalysis by anionic surfactants but is inhibited by cationic ones [74–79]. Fourth, the attack of anionic species on carboxylic esters is promoted by cationic surfactants but inhibited by anionic surfactants, and the reaction of neutral species with these substrates is little affected by ionic surfactants of any charge [76, 78–84]. Fifth, the addition of hydroxide ion to cationic dyes is subject to catalysis by cationic surfactants but to in-hibition by anionic ones [68]. This list can readily be expanded substantially, but the principal point in clear: the effects can be qualitatively understood in terms of electrostatic stabilization of the transition state, which possesses a charge or partial charge opposite to that of the micellar surface relative to the reactant state, generally an uncharged substrate in the micellar phase and a charged one free in aqueous solution. Not all reactions thus far examined fit neatly into this situation, and attention is directed to those that do not, but the substantial majority of them are in accord with these con-siderations.

The hydrophobic component of kinetic effects in micellar systems may be appreciated in terms of a few examples related to those just indicated. First, the degree of rate augmentation experienced on micellation for the hydrolysis of sodium alkyl sulfates increases very substantially as the length of the alkyl chain is increased from 10 to 18 carbon atoms [70]. Second, the acid-catalyzed hydrolysis of methyl orthobenzoate and methyl ortho-valerate is subject to catalysis by anionic surfactants; that for methyl orthoformate is not [76]. Third, the attack of N-myristoyl-L-histidine on p-nitrophenyl esters is markedly accelerated in the presence of micelles formed from hexadecyltrimethylammonium bromide while the attack of N-acetyl-L-histidine is not and, furthermore, the reaction of the former nucleophile with p-nitrophenyl hexanoate is much faster than that with the corresponding acetate [80]. Finally, rate and equilibrium constants for the addition of cyanide ion to N-alkyl pyridinium ions are markedly increased by cationic surfactants, and the magnitude of the change is accentuated with increasing chain length in the surfactants and with increasing chain length of the N-alkyl pyridinium ion [85]. Again, this list might be expanded a good deal, but the point is clear: kinetic effects in micellar systems are accentuated when hydrophobic interactions between substrate and surfactants are accentuated.

Among the various types of reactions that have received attention, the kinetics of a few of them in systems containing micelle-forming surfactants have been examined in sufficient detail to provide some general insight into the field, and we shall focus our attention on them. They include: (1) the hydrolysis of ortho esters and acetals, (2) nucleophilic reactions of carboxylic esters, (3) hydrolysis of phosphates and sulfates, (4) the fading of triphenyl-methyl dyes, (5) the addition of cyanide ion to pyridinium ions, and (6) nucleophilic aromatic substitution. From these investigations, a number of aspects of kinetic behavior in micellar systems have proved interesting: (1) the reaction site, (2) surfactant concentration-rate profiles, (3) substrate concentration-rate profiles, (4) surfactant structure, (5) substrate structure, (6) salt effects, and (7) effects of organic additives. These points are discussed sequentially below.

5.1 The Reaction Site

The localization of the site within the micelle phase at which bond changing reactions occur is of central importance for the understanding of the kinetics of these processes. Several lines of evidence strongly suggest that most reactions at least occur on the surface of the micelle, at or near the highly charged double layer that surrounds the hydrocarbon core and not within the hydrocarbon itself. It is true that no very clear line can be drawn between the micelle surface and the micelle interior, because the surface is rough and appreciable portions of the hydrocarbon chains are exposed to solvent, as noted before. We include the first two or three methylene groups of the hydrocarbon chains of the surfactant molecules as well as the hydrated charged groups (or polar nonionic head groups) as comprising the micellar surface. The distinction that we really wish to make is whether the reactions occur in a region of appreciable aqueous character. First, we have noted that there is substantial evidence that organic molecules possessing appreciable polar character are localized predominantly on the micellar surface. Second, it is a bit difficult to visualize reactions involving ionic species occurring readily within the hydrocarbonlike interior. Even if the interior is somewhat wet, we know of no evidence to suggest that ions from the bulk phase are included within this interior. Third, the rate of certain organic reactions is unaffected when one of the reactants is incorporated into a micelle and another is excluded from it. Such a reaction is that between anisylthioethane and iodine cyanide in the presence of sodium dodecyl sulfate [86]. Molecular sieve filtration experiments on Sephadex G-25 indicate that the former substrate is incorporated into the micelles but that the latter is not. Nevertheless, the rate of the reaction between them is unaffected by the surfactant. This is a difficult observation to rationalize if the anisylthioethane is not present on the micellar surface and, hence, accessible to the iodine cyanide. Fourth, the striking salt effects noted below are readily accommodated in terms of reactions occurring at the micellar surface but are more difficult to account for in terms of reactions occurring in the micellar interior. Finally, the reactive functionality of most substrates cannot project very far beyond the micellar surface because hydrophobic interactions between the micelle and substrate do seem to be important for the association between them. Taken as a whole, these results suggest that most reactions studied thus far in micellar systems do occur at the micellar surface and, furthermore, that the micellar surface is probably the principal habitat of the organic substrates (clearly, the principal site of occupation and site of reaction need not be the same). Exceptions do occur. For example, the observation that the basic hydrolysis of certain esters is retarded on their incorporation into micelles regardless of their charge suggests that these substrates are, in fact, hidden from solvent [83]. Menger and Portnoy have also concluded that p-nitrophenyl acetate and related esters are incorporated into the interior of micelles formed from lauric acid, although this interpretation is not the only possibility to account for the data [84].

5.2 Surfactant Concentration-Rate Profiles

In Figure 5, second-order rate constants for the hydrolysis of methyl orthobenzoate in dilute aqueous solutions of sodium dodecyl sulfate are plotted against the concentration of the surfactant [74]. The concentration-rate profile is multiphasic; below the cmc for the surfactant, the rate constants are independent of surfactant concentration. Above the cmc, the rate constants rise rapidly with increasing surfactant concentration, level off, and finally decrease with increasing concentration of this surfactant. At the optimal surfactant concentration, a rate augmentation of 85-fold is observed for this reaction. Although not all concentration-rate profiles show all these features, they do seem to be the general ones for reactions in micellar systems [74, 75, 78, 87–91]. Profiles of this type can be rationalized on the basis of (1) the necessity of micelles for catalysis, (2) adsorption of a progressively greater fraction of the substrate into the micellar phase until that fraction approaches unity with increasing surfactant concentration, and (3) inhibition of the micellar reaction by the counterions of the surfactant itself. In the case of methyl orthobenzoate hydrolysis in the presence of sodium dodecyl sulfate, it has been shown that this interpretation must be

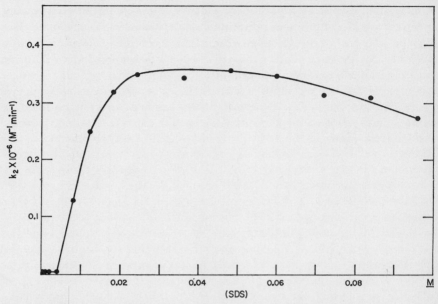

Figure 5   Second-order rate constants for hydrolysis of methyl orthobenzoate in aqueous solution at 25° plotted against the concentration of sodium dodecyl sulfate [R. B. Dunlap and E. H. Cordes, *J. Am. Chem. Soc.*, **90**, 4395 (1968)].

substantially correct [74]. Employing molecular sieve chromatography [86], the equilibrium constant for the association of substrate with surfactant was evaluated, $K_{assoc} = 73\ M^{-1}$. This value accounts quantitatively for the increase in rate constant with increasing surfactant concentration. That is, when the substrate is predicted to be 50 percent associated with the micellar phase on the basis of the equilibrium constant, about 50 percent of the maximum catalysis is experienced and so on. Furthermore, when the total concentration of sodium ion is maintained constant by the addition of the necessary quantities of inorganic sodium salts, the inhibition of the reaction at high surfactant concentrations disappears. A particularly nice example of this is shown for another reaction, the hydrolysis of *p*-nitrophenyl hexanoate in the presence of tetradecyltrimethylammonium chloride. If the chloride ion concentration is not maintained constant, a profile qualitatively similar to that of Figure 5 is obtained. However, when this concentration is maintained constant the profile of Figure 6 is obtained in which the inhibition has completely disappeared [78].

The general aspects of the profile in Figure 5 may change in two ways for other reactions. First, in some cases, there is evidence for the formation of small complexes between surfactant molecules and substrates at concentrations of the surfactant below the cmc and, in addition, for the induction of micelle formation by the substrate [73, 83]. In such instances, catalysis, or for that matter inhibition, occurs at surfactant concentrations lower than that for the cmc. Second, those reactions not involving ionic substrates probably do not experience strong inhibition by ions, and the inhibition at rather high surfactant concentrations is missing.

By neglecting the salt inhibition at high surfactant concentrations, it is possible to construct an approximate model which accounts semiquantitatively for the shape of the concentration-rate profile. Generally, this model will have the form [73, 83]:

$$M + S \underset{}{\overset{K}{\rightleftharpoons}} M \cdot S$$

$$\downarrow k_w \qquad\qquad \downarrow k_m \qquad\qquad (3)$$

$$\text{products} \qquad \text{products}$$

in which M is the micelle, S is the substrate, k_w and k_m are the rate constants for the reaction in the bulk phase and the micellar phase, respectively, and K is the equilibrium constant for association of the substrate with the micelle.

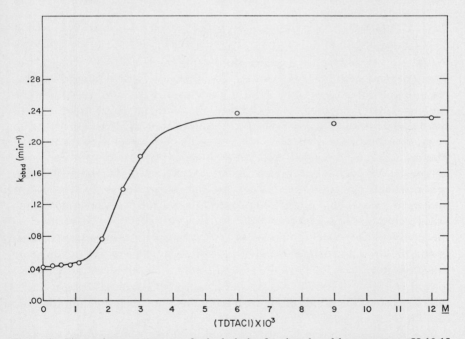

Figure 6   First-order rate constants for hydrolysis of p-nitrophenyl hexanoate at pH 10.15 plotted as a function of the concentration of tetradecyltrimethylammonium chloride at a constant total chloride ion concentration of 0.02 M [L. R. Romsted and E. H. Cordes, *J. Am. Chem. Soc.*, **90**, 4404 (1968)].

The concentration of micelles, C_m, is given by [73, 88]:

$$C_m = \frac{C_{total} - cmc}{N}$$ (4)

in which C_{total} is the total concentration of surfactant, and N is the number of surfactant molecules per micelle. Equation 4 requires that the concentration of surfactant be large with respect to that of substrate. This is the simplest model one can construct, but one that seems reasonably reliable provided no unusual effects occur. Bunton and his co-workers have provided a good example of the use of this model in studies on the kinetics of hydrolysis of 2,4-, and 2,6-dinitrophenyl phosphate hydrolysis [73]. The model predicts that plots of $1/(k_w - k_m)$ against $1/(C_m - cmc)$ will be linear, and in fact linear plots are found for this case except at rather low surfactant concentrations [73]. The deviations observed there may result from substrate-induced micellation of the surfactant or from the assumption that surfactant concentration is always large with respect to substrate concentration or from both. Although more sophisticated models can be constructed, these prove exceedingly cumbersome to employ and may, at any event, be themselves less than completely realistic.

A very interesting case of the dependence of rate on surfactant concentration is provided in the work of Gitler and Ochoa-Solano concerning the hydrolysis of p-nitrophenyl esters in the presence of mixed micelles of N^α-myristoyl-L-histidine (MirHis) and hexadecyltrimethylammonium bromide [80]. In this system, the MirHis is acylated by the substrates, and the cationic surfactant serves to solubilize the reactants and products. In Figure 7, first-order rate constants for the liberation of p-nitrophenol from several p-nitrophenyl esters are plotted as a function of the concentration of MirHis and as a function of the ratio of the concentration of MirHis to that of the cationic surfactant in the mixed micelles. Clearly, the reaction is linearly dependent on MirHis concentration only provided that there is relatively little cationic surfactant in the system and that the presence of the latter species inhibits the reaction. This can be understood in terms of a nonproductive binding of the substrates to the mixed micelles as an additional factor to those previously mentioned. The minimal kinetic scheme now must take the form:

$$S + Mi \underset{}{\overset{Ki}{\rightleftharpoons}} S \cdot Mi$$

$$S + Ma \xrightarrow{k_2} X\text{-}Ma + P_1$$ (5)

$$S\text{-}Ma + H_2O \xrightarrow{k_3} Ma + P_2$$

in which S is the substrate, Mi and Ma, respectively, the inactive and

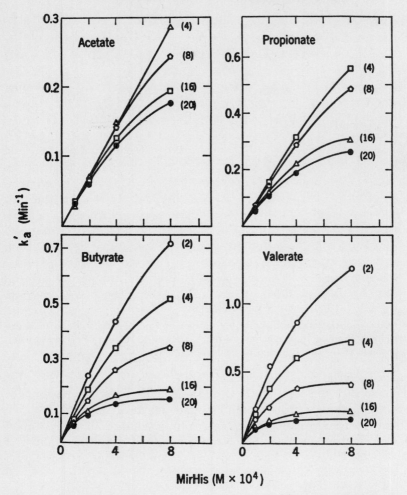

Figure 7 Pseudo-first-order rate constants for the liberation of p-nitrophenol in the reaction of mixed micelles of N^{α}-myristoylhistidine and cetyltrimethylammonium bromide (at the ratio indicated in parentheses) with p-nitrophenyl acetate (PNPA), propionate (PNPP), butyrate (PNPB), and valerate (PNPV) [C. Gitler and A. Ochoa-Solano, *J. Am. Chem. Soc.*, **90,** 5004 (1968)].

active positions within the mixed micelle, $S \cdot \mathrm{Mi}$ is a rapidly formed nonproductive complex, X-Ma is acylated MirHis in the mixed micelle, and P_1 and P_2 are p-nitrophenol and the acid corresponding to the ester employed, respectively. Analysis of the rate laws derived from this formulation yields values for the associated rate and equilibrium constants, some of which we refer to below. At the moment, it will suffice to recognize that acylation

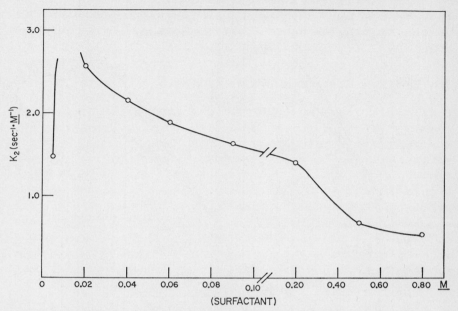

Figure 8   Second-order rate constants for the addition of cyanide ion to N-dodecyl 3-carbamoylpyridinium bromide as a function of the concentration of tetradecyltrimethylammonium bromide at 25° (P. Head and E. H. Cordes, unpublished data).

of the mixed micelle is very much more rapid than subsequent deacylation, as is the case with the reactions of these same esters with enzymes such as chymotrypsin and glyceraldehyde-3-phosphate dehydrogenase [92, 93].

In isolated cases, reaction kinetics have been studied in concentrated solutions of ionic surfactants. One example is provided by the addition of cyanide ion to N-dodecyl 3-carbamoylpyridinium bromide in a solution of n-tetradecyltrimethylammonium bromide [94].

Observed rate constants for this reaction are illustrated in Figure 8 as a function of surfactant concentration. At low concentrations the profile is typical; at higher concentration, the rates fall off. This must be partially the consequence of salt effects, but part of the effect may be due to changes in micellar structure. Over this concentration range, two transitions in structure do occur: a transition from spherical to rod-shaped micelles and one to an essentially liquid crystalline phase. Perhaps these transitions affect the reaction, but the effects are not sharp at the positions where the transitions occur.

5.3 Substrate Concentration-Rate Profiles

By analogy with other systems, including enzymatic ones, in which a complex is formed between two reactants prior to bond-changing reactions, one might expect that saturation of the micellar phase with increasing substrate concentrations would be observed. In certain cases at least, such behavior is found. In Figure 9, the first-order rate constants for hydrolysis of methyl orthobenzoate in the presence of $0.001\ M$ sodium dodecyl sulfate are plotted as a function of the concentration of the ortho ester [76]. The decreasing rate with increasing substrate concentration most likely represents saturation of the micellar phase with substrate. Thus as substrate concentration increases beyond the saturation point, an increasing fraction of the substrate must exist free in the solution. As this fraction approaches unity,

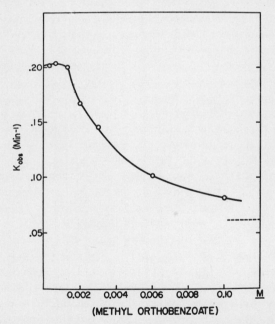

Figure 9   First-order rate constants for the hydrolysis of methyl orthobenzoate in the presence of $0.001\ M$ sodium dodecyl sulfate at 25° and pH 4.95 plotted as a function of the substrate concentration. The dotted line indicates the rate constant under these conditions in the absence of surfactant [M. T. A. Behme, J. G. Fullington, R. Noel, and E. H. Cordes, *J. Am. Chem. Soc.*, **87**, 266 (1965)].

the rate constant for the reaction must approach that for the reaction in purely aqueous solutions, as indeed it does.

5.4 Effect of Surfactant Structure

Perhaps the most dramatic change in surfactant structure, and the easiest to interpret in terms of influence on the kinetics of reactions in micellar phases, is change in the charge type of the head group. As developed briefly above, those reactions that are catalyzed by anionic surfactants are generally unaffected by nonionic ones and inhibited by those that are cationic and so on. In terms of electrostatic stabilization of the transition state relative to the reactant state as a crucial factor in establishing reaction rates in micellar systems, it is trivial to see why this should be so. Perhaps more subtle changes in structure will prove more challenging.

In maintaining the charge type of the surfactant constant there are two means of varying the surfactant structure: by changing the length or structure of the hydrocarbon chain and by changing the nature (but not the charge type) of the head group. Some information is available concerning both of these points. Quite generally, increasing the hydrophobic character of the surfactant increases its efficiency as a catalyst. Here we must distinguish two cases. In the first place, at equal concentrations of two surfactants, the more hydrophobic may appear to be the better catalyst (or inhibitor) simply because it has the greater affinity for the substrate. Thus with one surfactant, 80 % of the substrate may be associated with the micellar phase and with another at the same concentration, only 20 % may be so associated. It would not be surprising to find that the rate changes in the presence of the former surfactant are the larger. A nice example of such behavior has been provided by Bunton et al. [95]. These workers have observed that phenyl- and 2,4-dimethoxyphenylhexadecyltrimethylammonium bromides are more effective than hexadecyltrimethylammonium bromide as catalysts for the spontaneous hydrolysis of the dianions of 2,4- and 2,6-dinitrophenyl phosphate, the reactions of hydroxide and fluoride ions with p-nitrophenyl diphenyl phosphate, and for reaction of hydroxide ion with 2,4-dinitrochlorobenzene as a result of better micelle-substrate binding. In contrast, there is little difference in the rate of the micellar reactions in the presence of sufficient surfactant to adsorb substantially all of these substrates [95].

In the second place, differences in catalytic ability between related surfactants may persist even under conditions in which they are present at concentrations sufficiently high so that substantially all of the substrate is incorporated into the micellar phase in both cases. There are, in fact, several examples of both types of behavior. For example, the rate of increase of rate constants with increasing surfactant concentration for the hydrolysis of methyl orthobenzoate in the presence of sodium alkyl sulfates is more dramatic as the length of the hydrocarbon chain is extended [74]. Assuming that the equilibrium constant for association of this ortho ester with the micellar phase increases with increasing chain length, it follows that, at equal concentrations of surfactants, more of the substrate will be in the micellar phase derived from the more hydrophobic surfactant in accord with the observation that a greater rate change is observed. A number of related examples can be cited.

An interesting observation related to the latter point is the relation between rate constants and surfactant hydrophobicity under conditions of saturation of substrate with surfactant. Several examples in the literature indicate that more hydrophobic surfactants are also better catalysts under these conditions. Returning to methyl orthobenzoate hydrolysis in the presence of sodium alkyl sulfates for a moment, maximal rate increases for this reaction as a function of surfactant structure are collected in Table 6. Clearly, the longer the hydrocarbon chain of the surfactant, the better catalyst it becomes. These results are completely in accord with the observations that the rate of acid-catalyzed hydrolysis of micellated sodium alkyl sulfates increases with increasing chain length [70], that rate constants for reaction of p-nitrophenyl esters with hydroxide ion in the presence of alkyltrimethylammonium salts increase with increasing length of the alkyl group [78], and with the finding

TABLE 6 TEMPERATURE DEPENDENCE OF THE MAXIMAL RATE INCREASES ELICITED BY A SERIES OF SODIUM ALKYL SULFATES FOR METHYL ORTHOBENZOATE HYDROLYSIS IN AQUEOUS SOLUTION[a]

Sodium Alkyl Sulfate	Temperature (°C)	$k_2^0 \times 10^{-6}$[b] $(M^{-1}\,min^{-1})$	$k_2 \times 10^{-6}$[c] $(M^{-1}\,min^{-1})$	Maximum Rate Increase
Octyl	25.0	0.00502	0.0351 at 0.20 M	7.0
Decyl	40.0	0.0191	0.296 at 0.075 M	15.5
	32.5	0.0094	0.211 at 0.075 M	22.4
	25.0	0.00452	0.121 at 0.075 M	26.8
Dodecyl	40.0	0.0188	0.774 at 0.024 M	41.2
	32.5	0.0094	0.584 at 0.036 M	62.1
	25.0	0.00452	0.357 at 0.048 M	79.0
Tetradecyl	40.0	0.0168	1.37 at 0.030 M	81.5
	35.0	0.0122	1.06 at 0.015 M	86.9
	30.0	0.00864	0.793 at 0.020 M	91.8
Hexadecyl	45.0	0.0298	2.56 at 0.006 M	86

[a] R. B. Dunlap and E. H. Cordes, *J. Amer. Chem. Soc.*, **90,** 4395 (1968).
[b] Second-order rate constants in the absence of surfactant.
[c] Second-order rate constants for the reaction in the presence of the indicated concentrations of surfactants at which values maximum catalysis occurs.

that the rate and equilibrium constants for addition of cyanide ion to N-alkyl pyridinium ions in the presence of n-alkyltrimethylammonium salts with increasing hydrophobicity of the surfactant [85]. We return to the last of these reactions later. There are at least three plausible reasons why increasing the hydrophobic character of the surfactant might tend to make it a better catalyst for organic reactions under saturating conditions. First, the charge density of ionic groups at the surface may increase with increasing chain length, thus increasing the electric field at the reaction site. There is some evidence to indicate that this is the case, although it is difficult to know if the increased field is large enough to account for the observed differences in rate. Second, the exact positioning of the substrate with respect to the micellar surface may depend on the hydrocarbon moiety of the surfactant. At the moment, one is hard-pressed to examine this possibility directly. Third, the binding forces between the hydrophobic parts of the substrate and micelle may be employed to reduce the overall activation energy for the reaction. There is some evidence to suggest that this interesting possibility is an important factor, although further discussion must await our consideration of the effects of changes in substrate structure.

In terms of changing the nature of the head group without changing the charge type, only one thorough study has been carried out [75]. Examination of some 30 anionic surfactants as catalysts for methyl orthobenzoate hydrolysis reveals marked differences between them although all are active. Some of the most active appear to be exceptionally good catalysts indeed, and rate changes of several hundred fold might have been achieved had problems of surfactant availability and solubility not intervened. Although detailed interpretation of all the data is not possible, some generalizations can be drawn. First, moving the charged group in from the terminal position on the hydrocarbon chain toward the center diminishes catalytic efficiency. Second, dianions tend to be better catalysts than monoanions, and, third, sulfates are usually more effective catalysts than sulfonates.

5.5   Effect of Substrate Structure

There are two aspects of substrate structure that have proved interesting in terms of reaction kinetics in micellar systems: substrate hydrophobicity and polar substituents. Generally, increasing the hydrophobic character of the substrate increases the influence of the micellar phase on the velocity of the reaction, just as increasing the hydrophobicity of the surfactant tends to accentuate these effects. Typical examples are provided by surfactant-dependent ester hydrolysis [78, 80, 81, 84], sulfate ester hydrolysis [70], ortho ester hydrolysis [78], and addition of cyanide ion to pyridinium ions [85]. In some respects, studies in the last of these systems may be the most revealing; in Table 7 rate and equilibrium constants for the addition of cyanide to a series of N-alkyl 3-carbamoylpyridinium ions in the presence of a series of alkyltrimethylammonium ions are collected. With respect to both substrate and surfactant, increasing hydrophobic properties increases the reactivity and affinity of cyanide ion for the substrates. This system provides the best evidence for a point raised above: that hydrophobic

TABLE 7 EQUILIBRIUM CONSTANTS AND RATE CONSTANTS FOR DISSOCIATION OF 1,4-CYANIDE ADDUCTS OF A SERIES OF N-ALKYLPYRIDINIUM IONS IN AQUEOUS SOLUTIONS OF 0.02 M ALKYLTRIMETHYLAMMONIUM IONS AT 25°[a]

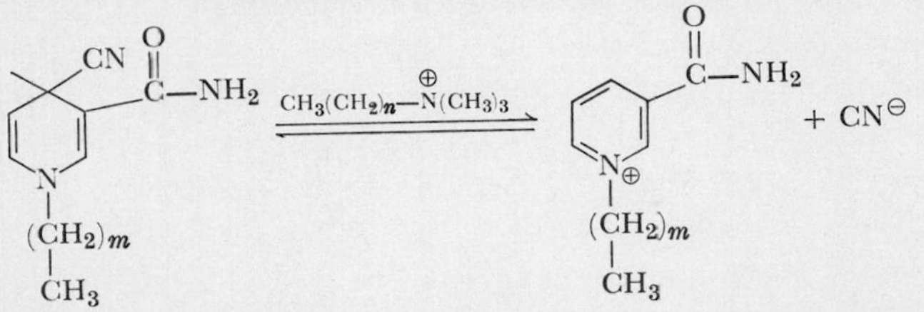

m \ n	9	11	13	15
		Equilibrium Constants (M)		
7				7.4×10^{-3}
9			1.9×10^{-3}	1.4×10^{-3}
11		5.5×10^{-4}		2.9×10^{-4}
13	3.1×10^{-3}	3.4×10^{-4}	2.8×10^{-4}	2.6×10^{-4}
15		2.2×10^{-4}		2.1×10^{-4}
		Rate Constants ($M^{-1} sec^{-1}$)		
7				0.21
9			1.08	1.35
11		2.46		5.77
13	0.28	5.84	6.57	10.4
15		6.38		13.3

[a] Data from reference 85.

interactions may be employed to reduce the activation energy [80]. At the concentrations employed in these studies, 0.02 M surfactant, the rate and equilibrium constants are maximal. Thus the substrates are associated substantially completely with the micellar phase. Furthermore, the properties of the micelles themselves should not be markedly dependent on the nature of the substrate, even following incorporation of the substrate into the micelles, since surfactant molecules outnumber substrate molecules about 200 to 1. One possible model within the framework of hydrophobic interaction contributions to activation energy is the following. In the course of the reaction process, a substrate possessing a polar headgroup is converted

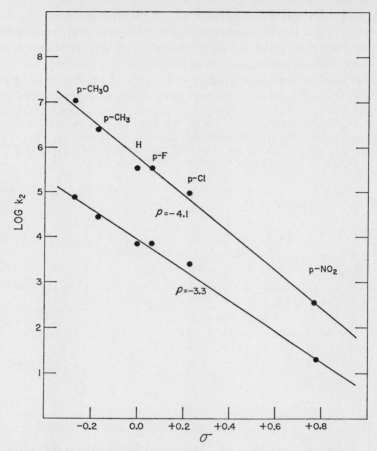

Figure 10   Logarithms of second-order rate constants (in units of $M^{-1}min^{-1}$) for the hydrolysis of a series of p-substituted benzaldehyde diethyl acetals in aqueous solution (lower line) and in the presence of sodium dodecyl sulfate (upper line) plotted against the Hammett substituent constants [R. B. Dunlap, G. A. Ghanim, and E. H. Cordes, *J. Phys. Chem.*, **73**, 1898 (1969)].

into a product in which this polar character is largely lost. As a result, the product molecule may occupy a somewhat different position with respect to the micellar structure than the reactant does. By increasing the hydrophobic character in the substrate, holding the headgroup constant, the position occupied by the substrate may approach that occupied by the product. Stated differently, the increasing hydrophobic interactions may drag the polar headgroup into the less agreeable environment that, in the product, will be occupied by an uncharged species. The hydrophobic interactions thus destabilize the reactant state with respect to the transition state and product state and contribute to the activation energy for the overall process. It is interesting to note that the rate and equilibrium constants for these reactions are 1000 to 20,000 times greater than for the same reactions in the absence of surfactants [85].

The effect of polar substituents on the rates of organic reaction in micellar phases appear to differ substantially from such effects in aqueous solution. In Figures 10 and 11, Hammett plots for the hydrolysis of a series of acetals

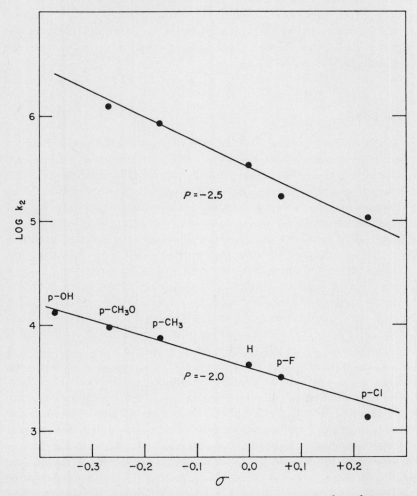

Figure 11 Logarithms of second-order rate constants (in units of $M^{-1}min^{-1}$) for the hydrolysis of a series of p-substituted methyl orthobenzoates in aqueous solution (lower line) and in the presence of sodium dodecyl sulfate (upper line) plotted against the Hammett substituent constants. Data from reference 75.

[77] and ortho esters [75] in water and in dilute solutions of sodium dodecyl sulfate are provided. Clearly, the reactions in the micellar phase are the more susceptible to polar effects. These results are not trivial to interpret. For the acetals at least, the indicated behavior may reflect changes in transition state structure as a function of the nature of the polar substituent [96]. It has been established through measurement of secondary deuterium isotope effects that increasing electron withdrawal in the polar substituent causes the transition state to assume more nearly carbonium ion geometry. This effect appears less pronounced for the ortho esters, however. Medium effects may be important as well although effects of polar substituents on the hydrolysis of the acetals in water and in 50 percent aqueous dioxane are not very different [77, 97].

5.6 Salt Effects

One of the striking aspects of the kinetics of organic reactions in micellar systems is their sensitivity to salt effects. Changes in the nature or concentration of electrolyte that would lead to barely detectable differences in rates of reactions in purely aqueous systems frequently cause differences of an order of magnitude or more for the same reactions in the presence of ionic surfactants. As an example, we have already noted the inhibition of surfactant-dependent reactions due to the counterion of the surfactant itself. Studies thus far clearly establish that hydrolysis of sulfate esters [70], hydrolysis of phosphate esters [70], hydrolysis of ortho esters [74], nucleophilic reactions of carboxylic esters [78, 82], and aromatic nucleophilic substitution reactions [87, 98] are all susceptible to marked inhibition by electrolytes. Two specific examples indicate the principal features of the inhibition. In Figure 12, the first-order rate constants for hydrolysis of p-nitrophenyl hexanoate at pH 10.15 in the presence of tetradecyltrimethyl-ammonium chloride are plotted as a function of the concentration of several monovalent anions. All anions studied are inhibitors, and 0.10 M concentrations of bromide and nitrate ions are sufficient to convert the surfactant-catalyzed reaction into a surfactant-inhibited one. As the hydrophobicity of the anion increases, and hence its tendency to associate with the micellar surface, the extent of inhibition is accentuated. Similar conclusions are evident from Figure 13, in which second-order rate constants for hydrolysis of methyl orthobenzoate in the presence of sodium dodecyl sulfate are plotted against the concentration of ammonium ions. Again, marked inhibition is observed and the inhibition increases as the hydrophobic character of the salt increases. These observations can be readily understood in terms of increasing the extent of charge neutralization of the micellar surface. To the

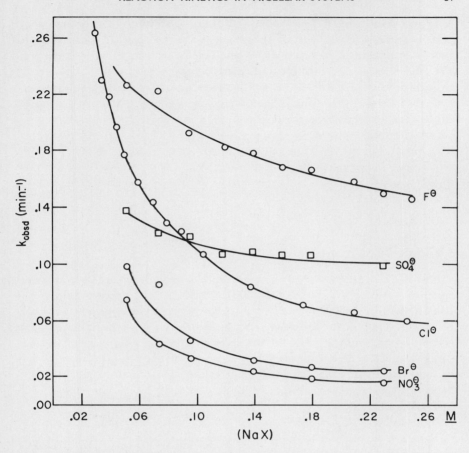

Figure 12   First-order rate constants for hydrolysis of p-nitrophenyl hexanoate in the presence of 0.009 M tetradecyltrimethylammonium chloride, pH 10.15, as a function of the concentration of several anions [L. R. Romsted and E. H. Cordes, *J. Am. Chem. Soc.*, **90,** 4404 (1968)].

extent that catalysis is dependent on electrostatic stabilization of the transition state with respect to the ground state, such charge neutralization must reduce the catalytic effect. In other cases, the salt inhibition may derive principally from the displacement of one reactant from the micellar surface by the electrolyte.

5.7   Effect of Organic Additives

Relatively little work has been accomplished concerning the effects of organic additives on the kinetics of surfactant-dependent reactions. Those conclusions that are available indicate that the effects are likely to be

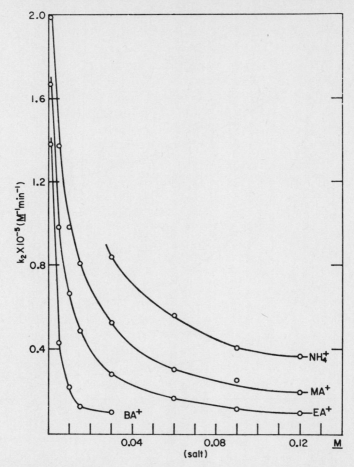

Figure 13   Second-order rate constants for methyl orthobenzoate hydrolysis in aqueous solutions containing 0.01 M sodium dodecyl sulfate plotted as a function of the concentration of unsubstituted, methyl, ethyl, and butylammonium ions [R. B. Dunlap and E. H. Cordes, *J. Am. Chem. Soc.*, **90**, 4395 (1968)].

complex and to differ from one system to another. For example, in Figure 14, second-order rate constants for the hydrolysis of methyl orthobenzoate in the presence of 0.01 M sodium dodecyl sulfate solution at 25° are collected as a function of the concentration of ethanol, n-butanol, n-heptanol, and dimethyldodecylphosphine oxide. As is apparent from the figure, the extent of inhibition of the surfactant-catalyzed reaction becomes greater as the concentration of alcohol is increased. Further, the effectiveness of the alcohol at any given concentration as an inhibitor increases with increasing length of the carbon chain of the alcohol molecule. As an example,

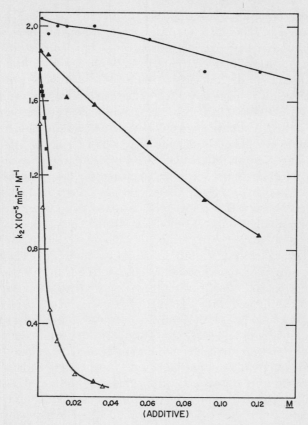

Figure 14 Second-order rate constants for the hydrolysis of methyl orthobenzoate in aqueous solution of 0.01 M sodium dodecyl sulfate at 25° plotted against the concentration of ethanol (●), n-butanol (▲), n-heptanol (■), and dimethyldodecyl phosphine oxide (△) [R. B. Dunlap and E. H. Cordes, *J. Phys. Chem.*, **73**, 361 (1969)].

n-decanol (not shown in the figure) at a concentration of $7.9 \times 10^{-5} M$ inhibits the reaction about twofold. The inhibition of the sodium dodecyl sulfate-dependent hydrolysis of methyl orthobenzoate by these nonionic additives may be the consequence of one or all of the following factors: (1) increase in the cmc of the anionic surfactant (particularly in the case of ethanol and n-butanol inhibition), thus lowering the concentration of micelles in the solution and hence the fraction of the ortho ester in the micellar phase; (2) displacement of methyl orthobenzoate from the micellar phase by the additives; or (3) lessening of the electrostatic stabilization of the transition state. There exists no basis for distinguishing between these factors or for assigning relative weights to them at this time. Our results do contrast with those for acid-catalyzed hydrolysis of micellated sodium dodecyl sulfate [71, 72]. The modest rate increases for this reaction observed upon the addition of long-chain alcohols or nonionic surfactants have been interpreted in terms of increasing effective charge resulting from a lowered dielectric constant at the micellar surface [72] or in terms of increasing charge density resulting from increased surfactant ionization [71]. Regardless of the correct explanation, it is clear that the influence of nonionic additives on sodium dodecyl sulfate hydrolysis and on sodium dodecyl sulfate-dependent methyl orthobenzoate hydrolysis is distinct.

The foregoing analysis of the various factors that influence rates and equilibria of organic reactions in micellar systems is based on consideration of electrostatic and hydrophobic forces. In some cases that have come to light, the effects observed are not easily assigned to these interactions and may result principally from local medium effects, or perhaps from electrostatic effects of a subtler sort. Although many reactions that have been probed are little affected or slightly inhibited by nonionic surfactants, some examples of catalysis by these species are known. Thus the reaction between 2,4-dinitrobenzene and anilines is subject to substantial catalysis by Igepal, dinonylphenol condensed with 24 units of ethylene oxide [98]. The same surfactant is also a catalyst for the hydrolysis of 2,4-dinitrophenyl sulfate [99]. A distinct example is provided by those additions of cyanide ion to N-alkyl pyridinium ions that are subject to promotion by the zwitterionic surfactant, dodecyldimethylammoniopropane sulfonate [85]. The maximal rate and equilibrium constant increases elicited by this surfactant are 71- and 5700-fold, respectively. Clearly, a more complete understanding of reactions in micellar systems requires the development of adequate explanations for these findings.

Finally, we may note that surfactants have potential uses as a means of altering the *course* of organic reactions as well as their rate and equilibrium constants. One fine example has been provided by the observation that the stereochemistry of the nitrous acid deamination of amines is affected upon micellation of the substrates [100].

Studies of reactions in micelles are in their infancy, and both continuation of work along current lines and development of new directions should yield interesting and significant findings. It will be of particular interest to examine the *course* of organic reactions in micellar systems in some detail. Preliminary results indicate that micelles exert some control over stereochemistry [100]; it seems likely that considerable advantage can be taken of this behavior. Moreover, it may prove possible to control the mode of ring closure in cases where more than one possibility exists. For example, a careful study of the course of solvolysis of farnesyl pyrophosphate seems an attractive system for investigation.

Both studies of kinetics and route for organic reactions in the presence of micelles should include use of surfactants that form distinctive structures, for example, biological surfactants. Some early results cited below suggest that lecithin and other phospholipids may be interesting in this respect. In addition, the bile salts form micelles of unusual structure (as noted above) and reactions on these asymmetric structures may prove to have unusual character. Finally, one can note the obvious extensions of current studies to a host of unstudied reactions. Included in this group are many reactions of significant biological importance, including polymerizations.

6 REACTION KINETICS IN LIPOSOMES AND POLYSOAPS

Our discussion up to this point has dealt with some of the similarities between the surfaces of micelles, membranes, and proteins. Moreover, we have noted that there are a substantial number of parallels between enzyme-catalyzed reactions and micelle-catalyzed ones. Hopefully, extension of the studies in micellar systems will reveal additional information concerning the susceptibility of organic reactions to catalysis by those medium effects, electrostatic effects, and effects of hydrophobic bonding that are characteristic of amphipathic surfaces. But there is certainly no compelling reason to restrict one's attention to micellar model systems; reaction kinetics in the presence of polyelectrolytes [101] and monolayers [102] have a number of features in common with those just described. Moreover, both polysoaps and liposomes show evidence of promise in this regard, although work pursued thus far is fragmentary at best.

The only study of polysoaps as catalysts for organic reactions that has come to our attention is that of Kirsh et al. who have employed poly-4-vinylpyridine partially alkylated with benzyl chloride as catalyst for p-nitrophenyl acetate hydrolysis [103]. These workers did observe a substantial catalytic effect, and the system mimicked the behavior of chymotrypsin in several respects, although the catalytic efficiency is markedly lower.

The observation that addition of cyanide ions is susceptible to catalysis by simple zwitterionic surfactants (p. 33) raises the possibility that biological zwitterionic surfactants such as lecithin might prove effective catalysts as well. Sonicated aqueous dispersions of egg lecithin form liposomal structures in which concentric bilayers of lecithin are separated from each other by aqueous layers a few water molecules thick (Fig. 2). Such preparations have indeed been found to be good catalysts for addition of cyanide to N-dodecyl-3-carbamoylpyridinium bromide [85, 104]. In Figure 15, the first-order rate constants and total change in optical density at 340 nm are plotted as a function of cyanide concentration for this reaction in the presence of $1.04 \times 10^{-4} M$ sonicated egg lecithin at 25°. Analysis of the

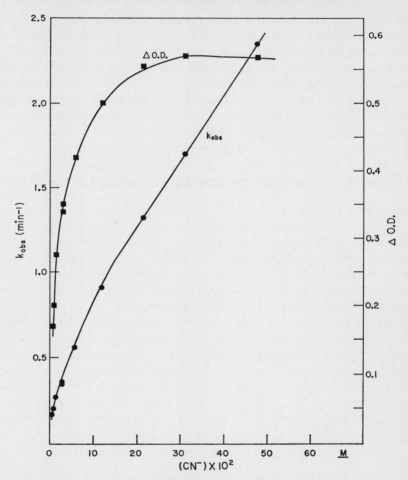

Figure 15   First-order rate constants (●) and total change in optical density (■) for addition of cyanide to *N*-dodecylnicotinamide at 25°, plotted as a function of cyanide ion concentration in the presence of 1.04×10^{-4} *M* sonicated liposome dispersions of ovolecithin in approximately 0.47 *M* NaCl and 0.093 *M* triethylamineammonium chloride buffer, pH 9.94 to 10.70. Initial concentrations of *N*-dodecylnicotinamide bromide and NaCN were 1.0×10^{-4} *M* and 0.033 *M*, respectively.

optical density data as well as visual inspection of that for the rate constants indicates that the reaction in the presence of liposomes is not a simple one. Specifically, the reaction is not accurately first-order in cyanide concentration, and the equilibrium constant is not independent of this variable. In view of the complexity of the liposomal structure, these observations are hardly surprising.

In Figure 16, first-order rate constants and total optical density changes for cyanide addition to N-dodecyl-3-carbamoylpyridinium bromide are plotted as a function of the concentration of egg lecithin. Although the considerations noted above make quantitative evaluation of this data difficult, it is clear that the lecithin preparation increases both the reactivity and the affinity of cyanide for the pyridinium ion and that the effects are quite marked at low concentrations. Sonicated dispersions of sphingomyelin and lysolecithin are also effective catalysts.

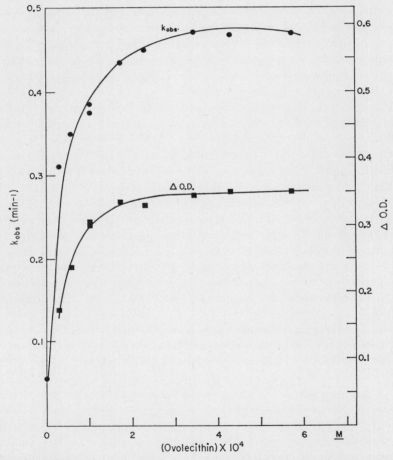

Figure 16   First-order rate constants (●) and total change in optical density (■) for addition of cyanide to N-dodecylnicotinamide at 25°, plotted as a function of ovolecithin concentration, present as a sonicated liposome dispersion in 0.47 M NaCl, 0.093 M triethyl-amineammonium chloride buffer, pH 10.04 ± 0.10. Initial concentrations of N-dodecyl-nicotinamide bromide and NaCN were 1.0 × 10^{-4} M and 0.033 M, respectively.

7 INTERACTIONS OF AMPHIPATHIC SURFACES WITH AMPHIPATHIC MOLECULES AND SURFACES

The first section of this chapter notes the fact that many biochemical phenomena involve interactions at amphipathic surfaces. Specifically, the chemistry of enzyme-catalyzed reactions may be viewed as reflecting the behavior of organic molecules at such surfaces, and we have reviewed in detail reaction kinetics in micellar systems as a model for these reactions. Two other classes of interactions also relate to matters of biochemical importance: interaction of amphipathic molecules with amphipathic surfaces and interaction of amphipathic surfaces with other such surfaces. Although detailed studies of suitable model systems that may shed light on the nature of these interactions and the consequences of them are currently lacking, it is appropriate to consider the importance of these interactions briefly.

The interaction of amphipathic surfaces, including those of proteins, membranes, and lipid aggregates, with related molecules and surfaces clearly underlies the following phenomena: the formation of proteins possessing quaternary structure and the cooperative response of such structures elicited by certain small organic molecules (effectors); the formation of the variety of lipoproteins that function in lipid transport and metabolism; the interaction of proteins with membranes and other forms of associated lipids and the functional behavior of proteins following such interaction; the interaction of proteins generally with surfactants; the interaction of enzymes with amphipathic substrates that include such species as long-chain fatty acids, phospholipids, sphingolipids, cholesterol, plus other steroids and tocopherols, and certain biosynthetic intermediates such as farnesyl pyrophosphate; and finally, the interaction of enzymes with amphipathic lipids that serve as cofactors.

From these various examples, the most pertinent ones deal with the effects of interaction between surfactants and phospholipids on the properties of enzymatic reactions. These studies have been viewed as parallel to those described in detail above for simple organic reactions. Essentially, they treat of the consequence of interactions at an amphipathic surface (that of the enzyme) on the behavior of other organic molecules (the substrates) at that surface. Thus they exhibit an order of complexity that is not found for nonenzymatic reactions.

A particularly nice example of the specificity of interaction of enzymes and surfactants and of the consequences of such interaction is provided by the studies of Tanaka and Sakamoto on a $(Na^+-K^+-Mg^{2+})$-activated ATPase isolated from bovine cerebral cortex [105]. Enzymes of this type were discovered in 1957 by Skou [106] and have subsequently been identified in a variety of tissues; they are particularly active in the membranes of excitable and secretory tissues, such as brain, kidney, and nerve [107]. It is generally considered that these enzymes are responsible for at least a major portion of the metabolically driven exit of sodium ions and entry of potassium ions across the plasma membrane. The enzymes have been isolated in active form only as complex lipoproteins, evidently as the protein moiety in association with a fragment of the plasma membrane with which it is associated in the living cell. Treatment that tends to disrupt these membranes too violently results in inactivation of the enzymes; this inactivation is frequently irreversible. These considerations suggest, and the studies of Tanaka and Sakamoto confirm, that interaction of the protein with components of the membrane is essential to activity *in vivo*.

When the $(Na^+-K^+-Mg^{2+})$-activated ATPase of bovine cerebral cortex is extracted by deoxycholate treatment, an inactive preparation is obtained that may be reactivated by a variety of surfactants [105]. Among the active surfactants, dialkyl phosphates are quite interesting. In Figure 17, the enzymatic activity of this ATPase is plotted as a function of the concentration of several such phosphates. In each case, the dialkyl phosphates increase enzymatic activity at low concentrations but then become inhibitory at higher concentrations. Both the maximal activity obtained and concentration at which maximal activity is elicited are sensitive functions of the alkyl chain length; didecyl phosphate is the most effective activator, but ditetradecyl phosphate is nearly inert in this capacity. Monoalkyl phosphates are also effective activators, and chain length again is an important determinant of the activation capacity of the surfactants; as above, optimal results are obtained through use of the decyl derivative.

A proper balance between surfactant hydrophobicity and hydrophilicity, while important, is not the sole determinant of activating efficiency toward the ATPase. The negative charge on the surfactant is also important; alkyl ammonium ions are completely ineffective and aliphatic alcohols only slightly effective. Fatty acids exhibit some activity, but mono-, di-, and triglycerides do not. Lecithin and phosphatidylethanolamine are only slightly effective, but phosphatidic acid, phosphatidylinositol, and phosphatidylserine markedly activate the ATPase. Thus minimal requirements for activation include a phosphate group together with a substantial hydrophobic moiety. The activation, hence, is quite specific. Lipid interaction with the enzyme must act to form a catalytically active constellation of amino acid side chains at one or more active sites; formation of a catalytically active center requires a precise interaction between the complexing entities.

These observations provide a rationale for a curious observation. Jain et al. have noted that reconstitution of ATP-mediated active transport of sodium and potassium ions, employing an enzyme from the synaptosomes of

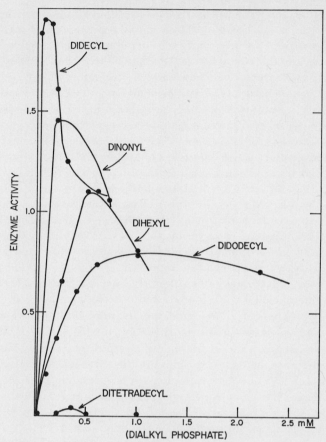

Figure 17 Activity of Na-K-Mg-dependent ATPase of beef cerebral cortex plotted as a function of the concentration of several dialkyl phosphates. From the data of reference 105.

rat brain, across a black lipid membrane constructed principally from oxidized cholesterol and alkanes, requires 15–30 parts per million (ppm) of didodecyl phosphate [108]. Neither lower nor higher concentrations are consistent with enzyme-mediated current flow across the membrane. Evidently, this behavior is closely related to that observed by Tanaka and Sakamoto; that is, enzyme activity depends on a specific interaction of the enzyme moiety with a surfactant. That surfactant must be the dialkyl phosphate; the remaining membrane components may serve as nothing more than a support.

A related case has been developed through the studies of Green, Fleischer, and their co-workers with the mitochondrial $D(-)$-β-hydroxybutyrate dehydrogenase isolated from beef heart. Like the membrane-bound ATPase, this enzyme is inactivated when it is removed from its mitochondrial locale [109, 110]. It, too, may be reactivated through the addition of phospholipids, and the reactivation is specific. Only micellar lecithin, in the presence of a sulfhydryl compound, is capable of effecting reactivation. Phosphatidyl-ethanolamine, cardiolipin, sphingomyelin, phosphatidylinositol, and phosphatidic acid are ineffective. These lipids do complex with the apodehydrogenase, but complexation does not generate enzymatic activity. The complexation and reactivation involving lecithin micelles are strongly temperature dependent and require 15 min for completion. The reconstituted enzyme appears to have catalytic properties identical to those of the enzyme in association with the original mitochondrial lipids.

A microbial dehydrogenase, flavin-dependent malate dehydrogenase from *Mycobacterium avium*, exhibits many of the same properties as the mitochondrial enzyme just discussed. Activity of this enzyme requires bacterial phospholipids; beef cardiolipin also supports the reaction [111]. Surfactants such as Tween 80 and deoxycholate are ineffective, and lecithin is moderately effective. Maximal activation with cardiolipin occurs at a lipid:protein weight ratio of only 1:40, strongly suggesting formation of a specific stoichiometric protein-lipid complex.

The biosynthesis of the cell wall lipopolysaccharide of gram-negative bacteria depends upon lipid cofactors of the cell wall itself [112]. The lipid requirement has been shown to result from demands of enzyme involved in the transfer of glucose and galactose groups in the biosynthetic process [113]:

$$\text{Glucose-deficient lipopolysaccharide} + \text{UDP-glucose} \xrightarrow{\text{phospholipid}}$$
$$\text{glucosyl-lipopolysaccharide} + \text{UDP} \quad (6)$$

$$\text{Glucosyl-lipopolysaccharide} + \text{UDP-galactose} \xrightarrow{\text{phospholipid}}$$
$$\text{galactosyl-glucosyl-lipopolysaccharide} + \text{UDP} \quad (7)$$

The phospholipid requirement of the enzyme catalyzing the latter of these reactions has been examined in some detail, and the requirement is again rather a specific one. Certain phospholipids were observed to form a complex with the enzyme but not with the lipopolysaccharide; others were observed to complex with the lipopolysaccharide but not with the enzyme. Neither class of phospholipid supports the reaction. Activity is generated only by those phospholipids, including certain phosphatidylethanolamines, that complex with both components, suggesting that the phospholipid forms a complex with both simultaneously in the course of galactose transfer. In Figure 18, the specificity of the reaction for the nature of the phosphatidylethanolamine is diagramed; clearly, a marked degree of specificity is observed as limited structural changes have marked effects on activating properties.

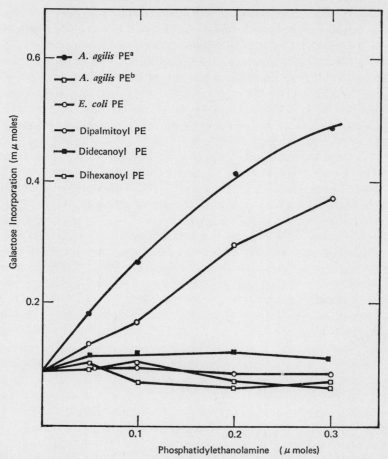

Figure 18   Effect of different phosphatidylethanolamines on the transferase reaction. Assays were carried out in the usual manner except that the type of phosphatidylethanolamine was varied as indicated in the figure. *A. agilis* PE^a refers to native phosphatidylethanolamine. *A. agilis* PE^b refers to phosphatidylethanolamine that has been subjected to catalytic hydrogenation. In each case the lipopolysaccharide-phospholipid mixture was heated and slowly, cooled prior to the addition of enzyme and UDP-galactose-^{14}C [A. Endo and L. Rothfield, *Biochemistry*, **8,** 3508 (1969)].

The foregoing examples establish that the interaction of enzymes and amphipathic species may have a profound effect on enzymatic activity itself. There is some evidence to suggest that such interactions may have subtler effects as well. For example, Nordlie and his associates have established that the polyfunctional glucose-6-phosphatase isolated from mammalian microsomes is susceptible to metabolic regulation by ATP and other nucleotides and, moreover, that the susceptibility of this enzyme to such regulation is markedly affected by surfactants [114]. Thus cetrimide, lysolecithin, and palmityl-SCoA all decrease the values of K_i for ATP and K_m for glucose-6-phosphate; the effect on K_i is substantially greater than that upon K_m. In consequence, each of these surfactants accentuates the degree of inhibition by a given concentration of ATP. Since the enzyme, being microsomal, exists *in vivo* in an amphipathic environment, these findings suggest that interaction of the enzyme with the membranes of the endoplasmic reticulum may have a profound influence on its susceptibility to metabolic control.

Interactions between proteins and amphipathic structures are certainly involved in a number of complex metabolic processes, including the clotting of blood [115] and the induction of lactose transport in *E. coli*. [116]. The molecular basis for the involvement of such interactions in blood clotting is understood in part. Bovine plasma prothrombin forms a complex with an equimolar mixture of phosphatidylcholine and phosphatidylserine. These entities together with activated factor X, factor V, and Ca^{2+} form an activated complex in which prothrombin is converted to thrombin. The latter protein does not form a similar complex and, hence, is released into the aqueous phase. Hence provision for the activation of additional molecules of prothrombin is made. Thus formation of the complex may both provide the necessary environment for the prothrombin-thrombin conversion and act to promote the conversion through displacement of the product molecule from the site of its synthesis.

These examples certainly suffice to establish the importance of interactions of enzymes and other proteins with amphipathic molecules and surfaces with respect to catalytic activity. Of a variety of additional examples that might be cited, one in which the substrate of the reaction is capable of assuming more than one micellar structure is particularly interesting.

Law and his co-workers have carried out an extensive investigation of the enzymatic reaction responsible for the synthesis of cyclopropane fatty acids [117]. It has been established that the substrates for the reaction are *cis*-unsaturated fatty acids that exist as components of phospholipids, usually phosphatidylethanolamines and that *S*-adenosylmethionine is the donor of the methylene group. Hence, the reaction may be formulated as shown on the following page.

Of particular interest here is the observation that the state of dispersion of the phospholipid is important for the action of the synthetase. The preferred method is dispersion of the phospholipid by a dialysis procedure rather than by mechanical or sonication techniques. This sort of dispersion yields products of very complicated structure; almost certainly, large aggregates are obtained consisting of ordered arrays of bimolecular leaflets formed through association of the hydrophobic portions of the molecules and separated by layers of water molecules into which the polar groups of the phospholipid

39. Average value found for nine different amphiphiles by P. Molyneux, C. T. Rhodes and J. Swarbrick, *Trans. Faraday Soc.*, **61**, 1043 (1965).
40. I. Langmuir, *J. Am. Chem. Soc.*, **39**, 1848 (1917).
41. C. Tanford, *J. Am. Chem. Soc.*, **84**, 4240 (1962).
42. J. T. Davies, *Trans. Faraday Soc.*, **48**, 1052 (1952).
43. D. S. Goodman, *J. Am. Chem. Soc.*, **80**, 3887 (1958).
44. K. Kinoshita, H. Ishikawa, and K. Shinoda, *Bull. Chem. Soc. (Japan)*, **31**, 1081 (1958).
45. A. Berger and I. Schechter, *Proc. Roy. Soc. London*, **Ser. B**, 189 (1970).
46. K. Shinoda, T. Nakagawa, B. Tamamushi, and T. Isemura, *Colloidal Surfactants* Academic, New York, 1963, p. 52.
47. M. Shinitzky, A. C. Dianoux, C. Gitler, and G. Weber, unpublished results.
48. M. T. Flanagan and S. Ainsworth, *Biochim. Biophys. Acta*, **168**, 16 (1968).
49. E. D. Goddard, C. A. J. Hoeve, and G. C. Benson, *J. Phys. Chem.*, **61**, 593 (1957).
50. K. Shigehara, *Bull. Chem. Soc. (Japan)*, **38**, 1700 (1965).
51. N. Muller and R. H. Birkhahn, *J. Phys. Chem.*, **71**, 957 (1967).
52. N. Muller and R. H. Birkhahn, *J. Phys. Chem.*, **72**, 583 (1968).
53. N. Muller and T. W. Johnson, *J. Phys. Chem.*, **73**, 2042 (1969).
54. P. Mukerjee, *J. Colloid Sci.*, **19**, 722 (1964).
55. D. Stigter, *J. Phys. Chem.*, **68**, 3603 (1964).
56. D. Stigter, *J. Colloid Interface Sci.*, **23**, 379 (1967).
57. J. Clifford, *Trans. Faraday Soc.*, **61**, 1276 (1965).
58. P. Mukerjee and A. Ray, *J. Phys. Chem.*, **70**, 2144 (1966).
59. S. Riegelman, N. A. Allawala, M. K. Hrenoff, and L. A. Strait, *J. Colloid Sci.*, **13**, 208 (1958).
60. J. C. Eriksson and G. Gillberg, *Acta Chem. Scand.*, **20**, 2019 (1966).
61. J. C. Eriksson and G. Gillberg, *Surface Chem.*, **1965**, 148.
62. M. Donbrow and C. T. Rhodes, *J. Chem. Soc.*, 6166 (1964).
63. M. Donbrow and C. T. Rhodes, *J. Pharm. Pharmacol.*, **18**, 424 (1966).
64. A. S. Waggoner, O. H. Griffith, and C. R. Christensen, *Proc. Natl. Acad. Sci.*, **57**, 1198 (1967).
65. E. J. Fendler and J. H. Fendler, *Advan. Phys. Org. Chem.*, **8**, 271 (1970).
66. P. H. Elworthy, A. T. Florence, and C. B. Macfarlane, *Solubilization by Surface-Active Agents and its Applications in Chemistry and the Biological Sciences*, Chapman Hall, London, 1968.
67. M. E. L. McBain and E. Hutchinson, *Solubilization*, Academic, New York, 1955.
68. E. F. J. Duynstee and E. Grunwald, *J. Am. Chem. Soc.*, **81**, 4540, 4542 (1959).
69. E. H. Cordes and R. B. Dunlap, *Acc. Chem. Res.*, **2**, 329 (1969).
70. J. L. Kurz, *J. Phys. Chem.*, **66**, 2240 (1962).
71. V. A. Motsavage and H. B. Kostenbauder, *J. Colloid Sci.*, **18**, 603 (1963).
72. B. W. Barry and E. Shotton, *J. Pharm. Pharmacol.*, **19**, 785 (1967).
73. C. A. Bunton, E. J. Fendler, L. Sepulveda, and K.-U. Yang, *J. Am. Chem. Soc.*, **90**, 5512 (1968).
74. R. B. Dunlap and E. H. Cordes, *J. Am. Chem. Soc.*, **90**, 4395 (1968).
75. R. B. Dunlap and E. H. Cordes, *J. Phys. Chem.*, **73**, 361 (1969).
76. M. T. A. Behme, J. G. Fullington, R. Noel, and E. H. Cordes, *J. Am. Chem. Soc.*, **87**, 266 (1965).
77. R. B. Dunlap, G. A. Ghanim, and E. H. Cordes, *J. Phys. Chem.*, **73**, 1898 (1969).
78. L. R. Romsted and E. H. Cordes, *J. Am. Chem. Soc.*, **90**, 4404 (1968).
79. E. F. J. Duynstee and E. Grunwald, *Tetrahedron*, **21**, 2401 (1965).
80. C. Gitler and A. Ochoa-Solano, *J. Amer. Chem. Soc.*, **90**, 5004 (1968).

81. A. Ochoa-Solano, G. Romero, and C. Gitler, *Science*, **156,** 1243 (1967).
82. P. Heitmann, *European J. Biochem.*, **5,** 305 (1968).
83. T. C. Bruice, J. Katzhendler, and L. R. Fedor, *J. Am. Chem. Soc.*, **90,** 1333 (1968).
84. F. M. Menger and C. E. Portnoy, *J. Am. Chem. Soc.*, **89,** 4698 (1967).
85. J. Baumrucker, M. Calzadilla, M. Centeno, G. Lehrmann, P. Lindquist, D. Dunham, M. Price, B. Sears, and E. H. Cordes, *J. Phys. Chem.*, **74,** 1152 (1970).
86. D. G. Herries, W. Bishop, and F. M. Richards, *J. Phys. Chem.*, **68,** 1842 (1964).
87. C. A. Bunton and L. Robinson, *J. Am. Chem. Soc.*, **90,** 5972 (1968).
88. M. T. A. Behme and E. H. Cordes, *J. Am. Chem. Soc.*, **87,** 260 (1965).
89. L. R. Cramer and J. C. Berg, *J. Phys. Chem.*, **72,** 3686 (1968).
90. C. A. Bunton and L. Robinson, *J. Org. Chem.*, **34,** 773 (1969).
91. C. A. Bunton and L. Robinson, *J. Org. Chem.*, **34,** 780 (1969).
92. F. J. Kézdy and M. L. Bender, *Biochemistry*, **1,** 1097 (1962).
93. M. T. A. Behme and E. H. Cordes, *J. Biol. Chem.*, **242,** 5500 (1967).
94. P. Head and E. H. Cordes, unpublished observations.
95. C. A. Bunton, L. Robinson, and L. Sepulveda, *J. Org. Chem.*, **35,** 108 (1970).
96. H. Bull, T. C. Pletcher, and E. H. Cordes, *Chem. Commun.*, **1970,** 527.
97. T. H. Fife and L. K. Jao, *J. Org. Chem.*, **30,** 1492 (1965).
98. C. A. Bunton and L. Robinson, *J. Am. Chem. Soc.*, **92,** 356 (1970).
99. E. J. Fendler, R. D. Liechti, and J. H. Fendler, *J. Org. Chem.*, **35,** 1658 (1970).
100. R. A. Moss and D. W. Reger, *J. Am. Chem. Soc.*, **91,** 7539 (1969).
101. H. Morawetz, *Advan. Catalysis*, **20,** 341 (1969).
102. G. L. Gaines, Jr., *Insoluble Monolayers at Liquid-gas Interfaces*, Interscience, New York, 1966, pp. 301ff.
103. Yu. E. Kirsh, V. A. Kabanov, and V. A. Kargin, *Proc. Acad. Sci. USSR*, **177,** 976 (1967).
104. M. Price and E. H. Cordes, unpublished observations.
105. R. Tanaka and T. Sakamoto, *Biochim. Biophys. Acta*, **193,** 384 (1969).
106. J. C. Skou, *Biochim. Biophys. Acta*, **23,** 394 (1957).
107. J. C. Skou, *Physiol. Rev.*, **45,** 596 (1965).
108. M. K. Jain, A. Strickholm, and E. H. Cordes, *Nature*, **222,** 871 (1969).
109. P. Jurtshuk, Jr., I. Sekuzu, and D. E. Green, *J. Biol. Chem.*, **238,** 1229 (1963).
110. S. Fleischer, G. Brierley, H. Klouwen, and D. B. Slautterbach, *J. Biol. Chem.*, **237,** 3264 (1962).
111. J. Tobari, *Biochem. Biophys. Res. Commun.*, **15,** 50 (1964).
112. L. Rothfield and B. L. Horecker, *Proc. Natl. Acad. Sci.*, **52,** 939 (1964).
113. A. Endo and L. Rothfield, *Biochemistry*, **8,** 3500, 3508 (1969).
114. R. E. Snoke and R. C. Nordlie, *Biochim. Biophys. Acta*, **139,** 190 (1967); R. C. Nordlie, T. L. Hanson, and P. T. Johns, *J. Biol. Chem.*, **242,** 4144 (1967); J. F. Soodsma and R. C. Nordlie, *Biochim. Biophys. Acta*, **191,** 636 (1969).
115. M. P. Esnouf and R. F. MacFarlane, *Advan. Enzymol.*, **36,** 255 (1967).
116. C. F. Fox, *Proc. Natl. Acad. Sci. U.S.*, **63,** 850 (1969).
117. J. H. Law, in *The Specificity of Cell Surfaces*, Davis and Warren, Eds., Prentice-Hall, Englewood Cliffs, N.J., 1967.
118. A. E. Chung and J. H. Law, *Biochemistry*, **3,** 967, 1989 (1964).
119. A. D. Bangham, *Advan. Lipid Res.*, **1,** 65 (1963).

The tides of fashion in protein chemistry are somewhat different from those in other fields. They never change; they merely become more inclusive. Practically every form of chemical interaction that is known to exist near atmospheric pressure and room temperature has been suggested for an important role in the structure of at least some protein. To this reviewer's knowledge none has ever been definitively ruled out. At this time the spotlight is on the apolar bond.

F. M. Richards, 1963

THE APOLAR BOND—A REEVALUATION

MICHAEL H. KLAPPER

Department of Chemistry, Division of Biochemistry, The Ohio State University, Columbus, Ohio

1 INTRODUCTION

The concept of the apolar bond was formalized by Kauzmann [1] who proposed that nonpolar amino acid side chains of a protein would tend to transfer into the nonaqueous environment provided by the protein interior for the same reasons that nonpolar solutes tend to transfer from water into organic solvents. Provided the analogy is correct, "burial" of such a side chain would be spontaneous, thus contributing energy for the stabilization of a particular three-dimensional polypeptide structure. As measured by the number of authors invoking the apolar bond, this proposal has been well accepted. This extensive, primary literature is covered in a number of excellent reviews [2, 3, 4, 5] obviating the Sisyphean attempt at an additional compendium. While seeking a point of view on which to base the chapter, the most consistent impression gained by this author was one of evanescence. The ambiguity surrounding the apolar bond is exemplified by the prefatory quotation from Richards [2]. Nine years after this statement, little change is apparent. The contribution of apolar bonding to the structural stabilization of proteins is universally accepted, but the theory remains incomplete, and the quantitative importance of this bond is yet to be established. This chapter, shaped by the uncertainties of its writer, therefore, does not contain extensive descriptions or categorizations of those experimental results for which the apolar bond may be invoked. The reader should regard this article as a heuristic discussion only which is primarily concerned with reexamining the premises upon which the concept of apolar bonding is based.

Since Kauzmann's original proposal is so intimately related to the theory of solutions, this topic is considered in most detail. We begin with a qualitative description of water, carbon tetrachloride (taken as the archetypal apolar solvent), and the solid water clathrate as solvents. An empirical theory of liquids is described, which can be used in the calculation of various solution properties. Finally, some consequences for a discussion of protein structure are mentioned. In recent years significant conceptual advances pertaining to the subject of this chapter have been made. These have not been compiled previously for presentation to biochemists, and maximum effort is placed on these subjects. Other, equally important material that has appeared to penetrate the biochemical literature is presented only briefly to avoid prolixity.

Before beginning, a note concerning nomenclature is required. Two models are currently available for discussing the interaction of apolar solutes and water, the "hydrophobic bond" of Kauzmann [1] and the "apolar bond" of Klotz [6]. Since we are interested in discussing theory without having to depend specifically on one or the other of these, a new generic designation was considered. This author could not construct an inherently better term, however, and settled upon the convention of using the term apolar bonding in a general sense. When a specific reference to one of the two models is required, the terms liquid-liquid partition (Kauzmann), and liquid-clathrate partition (Klotz) are employed.

2 SOLUTIONS OF APOLAR GASES

2.1 Liquid Solutions

The liquid-liquid partition model of the apolar bond is based on the partition of apolar solutes between water and an organic liquid. Kauzmann [1] proposed that the thermodynamic parameters which describe the transfer of small solutes between liquid phases may also apply to the "transfer" of protein side chains, with the aqueous environment and the protein interior replacing the water and the organic solvent. Were this proposal correct, it would be possible to compute the free energy contributed by apolar bonding to the stabilization of a protein's structure, because data on the partition of small molecules are obtained easily. There are major problems to be solved before such a computation is feasible. Three immediately obvious examples may be enumerated. Criteria with which to decide on the location, buried versus exposed, of any particular side chain must be established. The organic solvent that best mimics the protein interior must be chosen. The corrections required to describe an anchored side chain in terms of a freely moving solute must be determined. However, before attempting the solution of these, or other possible problems, the validity of the liquid-liquid partition model itself should be established.

If a bond with a large formation energy were to be considered, then it would be reasonable to assume that environmental perturbations would make minor contributions. For example, the formation energy of a carbon-carbon bond located in a protein structure should be approximated by the formation energy of a similar bond in a small molecule. In contrast, the formation energy of an apolar bond may be close to the magnitude of kT. Environmental effects should, therefore, have a major importance, and the direct application of a small molecule system would be problematical. It is intuitively obvious that the free energy of a methyl group in a protein side chain is not identical, except by chance, with the free energy of the methyl group in ethane. The premise behind the models of the apolar bond is that this free energy difference remains approximately constant when the two methyl groups are transferred from one similar environment to a second, provided the proper criteria of similarity have been established. To determine the validity of this premise we must be capable of describing the interactions between a molecule and its environment. We specifically require the physical bases of the thermodynamic parameters governing the formation of solutions; liquid solutions are discussed initially. Although the impetus of our discussion is a transfer process involving three components—two solvents and a solute—a liquid partition is thermodynamically equivalent to the vaporization of solute from one liquid, followed by condensation into the second. Because a two-component system is easier to analyze and because no difficulties are encountered on the reintroduction of the second solvent, we begin with the solution of apolar gases in liquids.

As noted by Eley more than 30 years ago [7] water appears, in comparison with organic liquids, to be an abnormally poor solvent for apolar gases. The apparently anomalous nature of water stands out in the entropies and enthalpies of the solution process. The Barclay-Butler rule [8, 9] is an empirical, hence, extrathermodynamic [10] relationship which states that

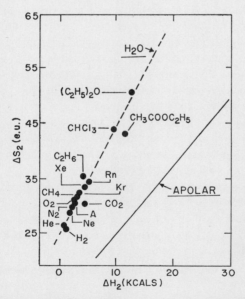

Figure 1   Barclay-Butler plot of apolar solutes in water at 25°C. The solid line is obtained from a best fit of the data for many gases in a number of organic solvents (95). This figure was adapted from Figure 3 of reference 11.

TABLE 1 ENTROPIES AND ENTHALPIES OF GAS VAPORIZATION FROM WATER AND CARBON TETRACHLORIDE AT $25°C^a$

Solute	ΔH_2^v(kcal/mole)^b		ΔS_2^v(cals/deg mole)^b	
	CCl_4	H_2O	CCl_4	H_2O
H_2	-1.26	1.28	11.7	26.0
N_2	-0.59	2.14	12.6	29.8
CO	-0.34	2.69	12.9	31.0
O_2	-0.01	2.99	13.4	31.3
CH_4	0.70	3.18	14.0	31.8
C_2H_6	2.71	4.43	16.8	35.4
C_3H_8	—	5.55^c	—	39.2^c
C_4H_{10}	—	6.28^c	—	42.8^c

^{a} All entropies and free energies presented in this chapter are on the unitary scale [21]. The standard state of the solute in the gas is taken as 1 atm and in solution as the hypothetical pure solute liquid that contains the solute-solvent interactions of the infinitely dilute solution.

^{b} All values taken from Frank and Evans [11] unless noted otherwise.

^{c} G. C. Kresheck et al. [12].

the vaporization entropy of any apolar gas dissolved in one particular liquid is directly proportional to the vaporization enthalpy of that gas. Frank and Evans [11] observed that the linear Barclay-Butler equations obtained with different apolar solvents are similar if not identical, and they proposed the use of a single relationship with all apolar solvents. The empirical equation derived from water solutions is also linear, but is distinctly different (Fig. 1). This is due to the more positive values of the thermodynamic parameters. A comparison of water and carbon tetrachloride solutions is presented in Table 1. Carbon tetrachloride has been taken as the archetypal apolar liquid, because more data have been collected with it than with any other liquid, save water. Because all apolar liquids behave similarly as solvents, the representation of the class by one of its members is permissible.

Other thermodynamic parameters also suggest that water is qualitatively different. The partial molar heat capacities of apolar gases are greater in water, as reflected in the much larger temperature dependence of the vaporization enthalpy [11]. The partial molar volumes of gases dissolved in water are lower than those of the same gases in organic solvents (Table 2). There is also evidence for significant differences between partial molar expansivities and partial molar compressibilities of gases dissolved in water

TABLE 2　THE　PARTIAL MOLAR VOLUMES OF GASES IN WATER AND CARBON TETRACHLORIDE AT 25°C

Gas	V_2(ml/mole)	
	$CCl_4{}^a$	H_2O
H_2	38	$<28^b$
Ar	44	—
O_2	45	31^c
N_2	53	35^b
CO	53	36^c
N_2O	47	—
CH_4	52	37^d
C_2H_4	61	—
C_2H_6	—	51^d
CF_4	80	—
C_3H_8	—	67^d
C_6H_6	—	83^d

^{a} Collected by Smith and Walkley [14].
^{b} Miller and Hildebrand [15].
^{c} Pierotti [16].
^{d} Masterton [17].

and other liquids [4]; the available data are, however, scanty for these two parameters, and little is known other than that they must be smaller in water.

Because of the hydrogen bonding potential of the water molecule, which can participate simultaneously in four hydrogen bonds, the structure of liquid water is distinctive. Although there is disagreement as to the details [19, 20], it is widely held that at room temperature water exists in an extensive, low-density network similar to, but not necessarily identical with ice I (melting point 0°C, 1 atm). This open structure in which the co-ordination number of each water molecule is not much larger than four is stabilized by hydrogen bonding. It is reasonable to assume, therefore, that the apparently unique properties of water as solvent arise from its structure, because apolar liquids cannot have a comparable hydrogen bond stabilized network. Theories based on a structural difference have, in fact, been proposed to explain the low solubilities of apolar gases in water. Two of these are described after the introduction of the following intellectual construction.

Because thermodynamic parameters are state functions, a physical or chemical process may be taken over any convenient path and divided into as many arbitrary steps as desired. During the course of the discussion, a number of different constructions are introduced to describe solute vaporization. It should be understood that no assumptions are inherent in these constructs and that no contradictions exist between them. For the first of these, solute vaporization will be taken to occur in two steps:

1. The solute is transferred from the liquid solution to the pure gas leaving a structured cavity behind.
2. The cavity then collapses with redistribution of the solution molecules.

If the two steps are designated with the subscripts a and b, then the enthalpy and entropy of solute vaporization may be expressed as:

$$\Delta H_2^v = \Delta H_a + \Delta H_b \tag{1}$$

$$\Delta S_2^v = \Delta S_a + \Delta S_b. \tag{2}$$

(Solution components are assigned the subscripts: 1 for the solvent, 2 for the solute, and larger digits for all other components.) With the use of this model qualitative statements concerning ΔH_2^v and ΔS_2^v may be made.

The vaporization enthalpy is considered first. Transfer of the solute to the gas phase in step a is an endothermic process, because of the intermolecular attraction between solute and solvent in the liquid phase. Solute-solute interactions in both the gas and the liquid may be ignored, and the mechanical work required for the solute expansion should be negligible. Cavity collapse, which occurs in step b, is an exothermic process. The molecules bordering the newly created hole acquire additional neighbors when the solvent is redistributed, resulting in a potential energy decrease. The net enthalpy, ΔH_2^v is thus the sum of positive and negative terms, and both its sign and magnitude are dependent on the absolute values of ΔH_a and ΔH_b. By inspection:

$$|\Delta H_a|/|\Delta H_b| > 1, \qquad \Delta H_2^v > 0$$
$$|\Delta H_a|/|\Delta H_b| < 1, \qquad \Delta H_2^v < 0.$$

It is also possible to predict that $|\Delta H_a|/|\Delta H_b|$ increases as the solute becomes larger. The work required to remove a molecule from the liquid interior to the vapor phase is proportional to the molecular volume, or for approximately spherical molecules the radius cubed; thus, ΔH_a is dependent on the cube of the solute radius. Formation of a cavity in the liquid without vaporization is equivalent to the transfer of one or more solvent molecules from the interior to the surface and is an enlargement of the liquid surface area. The work required to increase the surface is proportional to the cavity area, and thus to the square of the radius, again assuming a sphere. Because the collapse of an empty cavity in step b is the reverse process, ΔH_b is dependent on the square of the solute radius. We conclude that $|\Delta H_a|/|\Delta H_b|$ should increase as the solute size increases and that ΔH_2^v should become increasingly endothermic. This is the trend observed with both water and carbon tetrachloride (Table 1). With the latter solvent, a change in the sign of the enthalpy on passing from hydrogen to ethane clearly reflects that $\Delta H_2{}^v$ contains contributions from negative and positive terms.

The variation of the vaporization entropy, $\Delta S_2{}^v$, with solute size may also be predicted qualitatively. The entropy change of the first step, ΔS_a, contains three terms. Resolution of the solute from the solvent introduces a negative entropy of ideal mixing, $-\Delta S^M$, where:

$$\Delta S^M = -R \sum n_i \ln X_i \tag{3}$$

with n_i and X_i the number of moles and the mole fraction of the ith species in the mixture. Because of intermolecular interactions, there is a greater restraint on the rotational and vibrational movements of the solute in the liquid phase. This contributes a second term to ΔS_a. Finally, there is a greater volume restriction in the condensed phase because there is less available space for translation of the solute. The ideal mixing entropy is not dependent on the solute size [21] and is effectively eliminated if the entropy is calculated on the rational thermodynamic scale [1, 21], as has been done with the data presented in Table 1. The second term due to rotational and vibrational movements is negligible for apolar solutes. Justification of this statement is presented later. Thus the primary contribution to ΔS_a comes from release of the translational restriction; if the solute behaves ideally in the gas phase this entropy change is:

$$\Delta S_{\text{trans}} = R \ln \frac{V(g)}{v^f} \tag{4}$$

where $V(g)$ is the gas phase volume and v^f the effective or free volume [22] of the solute in the liquid solution. The hypothetical parameter v^f is the unoccupied space available to the solute. For a collection of hard spheres the free volume is approximated by the total volume of the collection less the volume actually occupied by all the spheres. In the well-known approximation of Van der Waals:

$$v^f = V - \frac{\pi}{12} \sum \delta_i{}^3 N_i \tag{5}$$

where δ_i is the diameter and N_i the number of the ith species. This equation is not correct for dense fluids in which the free volume is smaller due to excluded volume effects and intermolecular attractions [23]. If one uses equation 5 as a first approximation, however, it is apparent that v^f decreases with increasing solute size. Thus, ΔS_{trans} will become more positive (eq. 4), and since this is the only significant, size-dependent contribution to ΔS_a, this latter parameter also increases positively with an increase of the solute radius.

The entropy associated with cavity collapse, ΔS_b, is composed of two terms. The first is due to the additional restraints imposed on the rotational and vibrational movements of those molecules that, formerly on the cavity boundary, have an increased coordination number after the collapse. As before, we assume that this contribution is negligible. The origin of the second term becomes evident if the hole left by the vaporized solute is considered as a packing defect. Collapse is then the spread, or randomization of this defect throughout the liquid bulk; this introduces a positive term that should be larger in magnitude for larger cavities. Thus ΔS_b also becomes more positive with increasing solute radius. Since ΔS_2^v is the sum of two terms, both of which have a positive dependence on solute size, it has the same dependence; this is observed with both water and carbon tetrachloride solutions (Table 1).

The arguments just presented account for the observed trends in ΔS_2^v and ΔH_2^v, but cannot be used to predict the magnitudes of these parameters. Therefore additional logic is required to explain the differences between water and carbon tetrachloride. This is supplied in two different theories, one by Eley [7], the other by Frank, and co-workers [11, 24]. Both theories are based on the assumption that the low density, icelike network of water is the physical basis of this liquid's apparently unique solution properties. Eley proposed on the basis of the following argument that an apolar molecule dissolved in water would preferentially occupy the interstitial cavities within the network. At each lattice site a water molecule is coordinated to four neighbors by hydrogen bonds. Insertion of an apolar solute into a lattice site would require the prior destruction of these four bonds. Location of the solute in an available cavity would require no disruption of hydrogen bonds. Thus solution of apolar molecules into the interstitial sites would require less energy and would be favored. The more positive vaporization enthalpies and entropies of water solutions as compared with organic solvents can then be explained. Solute vaporization must be symmetric with solution formation. Since on dissolution in water available holes are filled, the opposite process, vaporization, must result in the creation of holes. In terms of the two-step construction we have been using, the cavities produced in removal of solute from water would not collapse, ΔH_b would be less negative, and the net vaporization enthalpy, ΔH_2^v, would be more endothermic. In an apolar solvent with no hydrogen-bonded network, there are no interstitial sites for preferential occupation, and the full contribution of ΔH_b would be seen.

The more positive vaporization entropies of water solutions may be explained in terms of the mixing entropy. The ideal mixing entropy (eq. 3), which for dilute binary solutions is:

$$\Delta S^M = -n_2 R \ln \left[\frac{N_2}{N_1 + N_2} \right] \tag{6}$$

is based on the assumption of complete interchangeability of solute and solvent at every site in the mixture. In an interstitial solution the solute does not occupy lattice positions, so that the mixing entropy becomes [7]:

$$\Delta S^M = -n_2 R \ln \left[\frac{N_2}{(N_1/X) - N_2} \right] \tag{7}$$

where X is the number of solvent molecules per interstitial site. With all reasonable values of X, the mixing entropy for vaporization calculated by equation 7 will be less negative than the ideal. Insofar as ΔS^M of water solutions is described by equation 7, and that of apolar liquids by equation 6, ΔS_2^v is more positive for the aqueous systems.

According to this argument ΔS^M is dependent on the manner in which the apolar solute is packed into the solution. If only a fraction of the total volume is accessible to the solute, then a correction is required to account for the nonideal mixing. A second packing contribution to the vaporization entropy is due to ΔS_{trans} (eq. 4). For dilute solutions the free volume (eq. 5) may be approximated by:

$$v^f = \left[V_1 - \frac{\pi}{12} \delta_1^{\,3} N_1 \right] - f(\delta_2, N_2) \tag{8}$$

$$= v_1^{\,f} - f(\delta_2, N_2)$$

where it is assumed that the dilute solution and pure solvent volumes are approximately the same, so that the solute contribution to v^f may be treated as an additive function of its radius. When equations 4 and 8 are combined, it is clear that ΔS_2^v becomes more positive for smaller values of the solvent-free volume. Estimates of this parameter obtained from the properties of pure liquids [22] indicate that the free volume of water is smaller than those of apolar solvents. Referring once again to the formal thermodynamic scheme of equations 1 and 2, Eley has proposed that the vaporization entropy from a water solution is more positive because of the more positive value of ΔS_a, which is, in turn, dependent on ΔS^M and ΔS_{trans}.

Frank and co-workers [11, 24], have suggested an alternative explanation for the apparently anomalous behavior of water as a solvent. These authors proposed an equilibrium between water aggregates, structured in an icelike lattice, and nonstructured water in which the hydrogen-bonded network is destroyed. The transition from ordered to nonordered water should occur with a positive enthalpy from disruption of the hydrogen bonds in the lattice. It was then assumed that the position of the equilibrium between the two water species is shifted toward the structured form by apolar solutes. The promotion of the icelike aggregates was attributed to the solute's ability to protect the cooperatively hydrogen-bonded network from adverse energy fluctuations, thereby lengthening the aggregate lifetime [18, 24, 25]. If mixing solute and water increases the amount of icelike water, then the reverse process of solute vaporization must decrease the amount of this species. This shift in equilibrium would introduce positive contributions to both the vaporization enthalpy and entropy. In comparison, therefore, apolar liquid solutions, in which there are no hydrogen-bond stabilized networks, would exhibit more negative vaporization parameters. In terms of equations 1 and 2, the theory of Frank and co-workers explains the greater enthalpies and entropies obtained with water solutions as arising from the more positive contributions of ΔH_b and ΔS_b. Although the theories of Eley, and Frank et al. assume different mechanisms, they are not mutually exclusive. The postulate that an apolar solute promotes water structure requires that the solute is preferentially located together with the structured water, that is, an interstitial solution. Thus the mixing entropy argument of Eley is also an inherent part of the structural promotion model, an interrelatedness that has been noted by Frank and co-workers [26, 27]. Both theories, moreover, attribute the greater vaporization enthalpies to the same step, cavity collapse, albeit with entirely different premises.

Vaporization free energies calculated from the data of Table 1 are collected in Table 3. Apolar gases, relatively insoluble in carbon tetrachloride, are even less soluble in water as seen by the negative Gibbs free energies of transfer from water to carbon tetrachloride (Table 3). The two-species equilibrium model of Frank and co-workers does not predict this free energy difference [26, 27] as was shown with an argument first given by Frank and Franks [27] and repeated here in a slightly altered form.

Assume that liquid water is composed of two species which are characterized by the different molar volumes V_k° and V_l°, and which are in an equilibrium

TABLE 3 FREE ENERGIES OF GAS VAPORIZATION FROM WATER AND CARBON TETRACHLORIDE AT 25°C

Solute	ΔG_2^v(kcal/mole)		ΔG_{tr}(kcal/mole) $H_2O \rightleftharpoons CCl_4$
	CCl_4	H_2O	
H_2	−4.75	−6.47	−1.72
N_2	−4.35	−6.74	−2.40
CO	−4.19	−6.55	−2.37
O_2	−4.01	−6.34	−2.34
CH_4	−3.47	−6.30	−2.83
C_2H_6	−2.30	−6.12	−3.83
C_3H_8	—	−6.14	—
C_4H_{10}	—	−6.48	—

described by:

$$K = \frac{N_l}{N_k} = \frac{1 - z}{z} \tag{9}$$

where N_k and N_l are the number of molecules of each species and z is the fraction in the k form. We now introduce a second, purely formal thermodynamic scheme. We consider vaporization from the two solutions of an apolar solute dissolved in pure species k or pure species l. The vaporization is performed in two steps. In the first, solute and solvent are isothermally separated into two pure liquids. The molar volume of the liquid solute, V_2°, may be assigned any convenient value, since in the second step of the overall process the liquid is vaporized. A value is chosen such that the free volume of the liquid solute $v_2{}^f$ is identical with $v_l{}^f$, the free volume of the l water species. In the second step of the overall process the liquid solute is vaporized. The entropy of liquid mixing may be expressed in terms of the liquid-free volumes [22]. For the mixture of liquid solute with k or l water the proper expressions are:

$$-\frac{\Delta S_{2,k}^M}{R} = n_k \ln\left[\frac{n_k v_k{}^f}{n_k v_k{}^f + n_{2,k} v_2{}^f}\right] + n_{2,k} \ln\left[\frac{n_{2,k} v_2{}^f}{n_k v_k{}^f + n_{2,k} v_2{}^f}\right] \tag{10}$$

$$-\frac{\Delta S_{2,l}^M}{R} = n_l \ln\left[\frac{n_l v_l{}^f}{n_l v_l{}^f + n_{2,l} v_2{}^f}\right] + n_{2,l} \ln\left[\frac{n_{2,l} v_2{}^f}{n_l v_l{}^f + n_{2,l} v_2{}^f}\right]. \tag{11}$$

$\Delta S_{2,k}^M$ and $\Delta S_{2,l}^M$ are the mixing entropies for liquid solute plus pure k and pure l water, respectively. The number of moles of solute in the two solutions is designated by $n_{2,k}$ and $n_{2,l}$; n_k and n_l are the moles of solvent in each solution.

The physically real solution of an apolar solute in water is assumed to contain n_2 moles of solute dissolved unequally in the k and l species, which in turn are dissolved into each other. Therefore:

$$n_2 = n_{2,k} + n_{2,l}$$

$$y = \frac{n_{2,k}}{n_2} \tag{12}$$

$$1 - y = \frac{n_{2,l}}{n_2}$$

with y the fraction of solute dissolved in the k water. To constitute the real solution, the two hypothetical solutions of solute in pure k and pure l are mixed according to equations 9 and 12. Assuming that the two water species obey independent mixing laws with respect to the solute and each other, the mixing entropy for formation of the real solution is:

$$\Delta S^M = \Delta S^M_{2,k} + \Delta S^M_{2,l} + \Delta S^M_{k,l} \tag{13}$$

with:

$$-\frac{\Delta S^M_{k,l}}{R} = n_k \ln\left[\frac{n_k v_k^{f}}{n_k v_k^{f} + n_l v_l^{f}}\right] + n_l \ln\left[\frac{n_l v_l^{f}}{n_k v_k^{f} + n_l v_l^{f}}\right] \tag{14}$$

The Gibbs free energy for the solution composed of the apolar solute and the two water species is the sum of the free energies of each component plus the mixing free energy, $T\Delta S^M$:

$$G_{\text{soln}} = n_{\text{H}_2\text{O}}[zG^\circ_k + (1-z)G^\circ_l] + n_2[yG^*_k + (1-y)G^*_l] - T\,\Delta S^M. \tag{15}$$

G°_k, G°_l, G^*_k, and G^*_l are the free energies of the k and l water species and of the solute dissolved in k and l. Introducing the definitions:

$$\begin{aligned}\Delta G^\circ_1 &= G^\circ_k - G^\circ_l\\ \Delta G^*_2 &= G^*_k - G^*_l\end{aligned} \tag{16}$$

equation 15 becomes:

$$G_{\text{soln}} = n_{\text{H}_2\text{O}}[z\,\Delta G^\circ_1 + G^\circ_l] + n_2[y\,\Delta G^*_2 + G^*_l] - T\,\Delta S^M. \tag{17}$$

Differentiation with respect to the three variables z, y and n_2 yields:

$$\left(\frac{\partial G_{\text{soln}}}{\partial z}\right)_{n_{\text{H}_2\text{O}},n_2,y,T} = n_{\text{H}_2\text{O}}\,\Delta G^\circ_1 - T\left(\frac{\partial \Delta S^M}{\partial z}\right)_{n_{\text{H}_2\text{O}},n_2,y,T} = 0 \tag{18}$$

$$\left(\frac{\partial G_{\text{soln}}}{\partial y}\right)_{n_{\text{H}_2\text{O}},n_2,z,T} = n_2\,\Delta G^*_2 - T\left(\frac{\partial \Delta S^M}{\partial y}\right)_{n_{\text{H}_2\text{O}},n_2,z,T} = 0 \tag{19}$$

$$\left(\frac{\partial G_{\text{soln}}}{\partial n_2}\right)_{n_{\text{H}_2\text{O}},z,y,T} = G^*_1 + y\,\Delta G^*_2 - T\left(\frac{\partial \Delta S^M}{\partial n_2}\right)_{n_{\text{H}_2\text{O}},z,y,T} \tag{20}$$

At equilibrium the solution free energy is a minimum. By imposing an equilibrium condition at a fixed-solute concentration, equations 18 and 19 may be set to zero. Combining equations 19 and 20 eliminates ΔG^*_2:

$$\left(\frac{\partial G_{\text{soln}}}{\partial n_2}\right)_{n_{\text{H}_2\text{O}},z,y,T} = G^*_1 + \frac{yT}{n_2}\left(\frac{\partial \Delta S^M}{\partial y}\right) - \left(\frac{\partial \Delta S^M}{\partial n_2}\right). \tag{21}$$

To obtain the difference on the right side of equation 21, equations 10, 11, 13, and 14 are combined and then differentiated with respect to y or n_2.

$$\begin{aligned}\left(\frac{\partial \Delta S^M}{\partial y}\right)_{n_{\text{H}_2\text{O}},n_2 z,T} = -n_2 R\Bigg\{ & n_l\left[\frac{v_2^{f} - v_l^{f}}{n_l v_l^{f} + (1-y)n_2 v_2^{f}}\right] - n_k\left[\frac{v_2^{f} - v_l^{f}}{n_k v_k^{f} + yn_2 v_2^{f}}\right]\\ & + \ln\left[\frac{yn_2 v_2^{f}}{n_k v_k^{f} + yn_2 v_2^{f}}\right] - \ln\left[\frac{(1-y)n_2 v_2^{f}}{n_l v_l^{f} + (1-y)n_2 v_2^{f}}\right]\Bigg\}\end{aligned} \tag{22}$$

$$\left(\frac{\partial \Delta S^M}{\partial n_2}\right)_{n_{H_2O}, y, z, T} = -R\left\{(y-1)n_l\left[\frac{v_2{}^f - v_l{}^f}{n_l v_l{}^f + (1-y)n_2 v_2{}^f}\right]\right.$$

$$- y n_k\left[\frac{v_2{}^f - v_k{}^f}{n_k v_k{}^f + y n_2 v_2{}^f}\right] + y \ln\left[\frac{y_2 n_2 v_2{}^f}{n_k v_k{}^f + y n_2 v_2{}^f}\right]$$

$$+ \left.(1-y)\ln\left[\frac{(1-y)n_2 v_2{}^f}{n_l v_l{}^f + (1-y)n_2 v_2{}^f}\right]\right\}. \tag{23}$$

When 22 and 23 are substituted into equation 21:

$$\left(\frac{\partial G_{soln}}{\partial n_2}\right)_{n_{H_2O}, z, y, T} = G_1^* - RT\left\{\frac{n_l(v_2{}^f - v_l{}^f)}{n_l v_l{}^f + (1-y)n_2 v_2{}^f}\right.$$

$$- \left.\ln\left[\frac{(1-y)n_2 v_2{}^f}{n_l v_l{}^f + (1-y)n_2 v_2{}^f}\right]\right\}. \tag{24}$$

Because solutions of apolar gases in water are very dilute, the following approximation is valid at saturation:

$$\left(\frac{\partial G_{soln}}{\partial n_2}\right) \approx G_2^\circ \tag{25}$$

where G_2° is the free energy of the pure liquid solute. Inserting equation 25 into 24 and rearranging yields:

$$G_2^\circ = G_1^* - RT\left[\frac{v_2{}^f - v_b{}^f}{v_b{}^f} - \ln\left(\frac{v_2{}^f}{v_b{}^f}\cdot\frac{1-y}{1-z}\right) + \ln\left(\frac{n_2}{n_{H_2O}}\right)\right]. \tag{26}$$

The unitary free energy for resolving a dilute binary solution into its pure components is:

$$\Delta G = -RT\ln\frac{n_2}{n_{H_2O}} \tag{27}$$

so that:

$$\Delta G = G_2^\circ - G_1^* + RT\left[\frac{v_2{}^f - v_l{}^f}{v_l{}^f} - \ln\left(\frac{v_2{}^f}{v_l{}^f}\cdot\frac{1-y}{1-z}\right)\right]. \tag{28}$$

Equation 28 is the free energy for transfering 1 mole of solute from the water solution to the pure liquid characterized by the free volume $v_2{}^f$. Since this quantity was initially assumed equal to $v_l{}^f$ equation 28 becomes:

$$\Delta G = G_2^\circ - G_1^* - RT\ln\frac{1-y}{1-z}. \tag{29}$$

To obtain the unitary free energy $\Delta G_2{}^v$ for vaporization of the solute from the aqueous solution, the free energy of converting the pure liquid solute into a gas $\Delta G_{2,0}^v$ must be added to equation 29. Hence:

$$\Delta G_2{}^v = G_2^\circ - G_1^* + \Delta G_{2,0}^v - RT \ln \frac{1-y}{1-z}. \tag{30}$$

As derived for dilute solutions, $\Delta G_2{}^v$ in the form of equation 30 does not contain a contribution due to the conversion of water from the l to the k species. Thus while z, a measure of the position of the water equilibrium in the aqueous solution, affects the solute solubility, the equilibrium shift that according to this theory occurs whenever $z \neq y$ does not appear in the net vaporization free energy. To understand how this occurs we need to derive the solute vaporization enthalpy $\Delta H_2{}^v$. Equations 18–20 may be rearranged and differentiated with respect to temperature:

$$-\frac{\Delta H_1^\circ}{T^2} = \frac{1}{n_{H_2O}} \frac{\partial}{\partial T} \left(\frac{\partial \Delta S^M}{\partial z} \right)_{n_{H_2O}, n_2, y, T} \tag{31}$$

$$-\frac{\Delta H_2}{T^2} = \frac{1}{n_2} \frac{\partial}{\partial T} \left(\frac{\partial \Delta S^M}{\partial y} \right)_{n_{H_2O}, n_2, z, T} \tag{32}$$

$$-\frac{H_2}{T^2} = -\frac{H_1^*}{T^2} - y\frac{\Delta H_2^*}{T^2} + \frac{\partial}{\partial T} \left(\frac{\partial \Delta S^M}{\partial n_2} \right)_{n_{H_2O}, z, y, T}. \tag{33}$$

Similarly with equation 30:

$$\Delta H_2{}^v = H_2^\circ - H_1^* + \Delta H_{2,0}^v - RT^2 \left[\frac{1}{1-y}\left(\frac{\partial y}{\partial T}\right) - \frac{1}{1-z}\left(\frac{\partial z}{\partial T}\right) \right] \tag{34}$$

Combining equations 10–14 with equations 31–34 yields:

$$\Delta H_2{}^v = H_2^\circ - H_b^* + \Delta H_{2,0}^v - y\,\Delta H_2^* - (y-z)\,\Delta H_1^\circ. \tag{35}$$

Moreover:

$$\Delta S_2{}^v = -\frac{\Delta G_2{}^v - \Delta H_2{}^v}{T} \tag{36}$$

so that the entropy of solute vaporization is:

$$\Delta S_2{}^v = S_2^\circ - S_1^* + \Delta S_{2,o}^v + R \ln\left(\frac{1-y}{1-z}\right) - y\frac{\Delta H_2^*}{T} - (y-z)\frac{\Delta H_1^\circ}{T}. \tag{37}$$

The postulated shift in water species, which occurs when $z \neq y$, is reflected in both the vaporization enthalpy and entropy as shown by the inclusion of ΔH_1° in equations 35 and 37. The absence of a corresponding term in the vaporization free energy is due to a complete enthalpy and entropy compensation; that is, if a shift of the water species equilibrium does occur during the solution process, then the contribution of this process to $\Delta G_2{}^v$ is zero.

In the argument used, no physical descriptions of the two species are required other than the assigned molar volumes. Since this is a disposable parameter, equations 30, 35, and 37 are valid for any two species model regardless of the physical differences postulated. Furthermore, the forms of these three equations are not altered when additional water species are considered. For a multispecies model equations 9 and 12 may be replaced by:

$$K = \frac{\sum N_i}{N_k} = \frac{1 - z}{z} \tag{9'}$$

$$\sum N_i = N_l + N_m + N_n + \cdots$$

$$n_2 = n_{2,k} + \sum n_{2,i}$$

$$1 - y = \frac{\sum n_{2,i}}{n_2} \tag{12'}$$

Equation 30 then becomes:

$$\Delta G_2^v = G_2^\circ - \sum G_i^* + \Delta G_{2,0}^v - RT \ln \left(\frac{1 - y}{1 - z} \right) \tag{30'}$$

and similar changes are introduced into the enthalpy and entropy expressions. We conclude that if the equilibrium positions between any number of water species are shifted by the solution of an apolar solute, the solution entropy and enthalpy would be sensitive to this physical event, but the free energy, or solute solubility would not.

2.2 Water Clathrate Solutions

Eley, and Frank et al. assumed that apolar solutes in water occupy cavities within or connected to hydrogen bonded, icelike networks. The cavities in the ice I (m.p. 0°C at 1 atm) lattice are too small to accommodate most solutes, excluding this species as the structured water required in either theory. Water does form, however, a variety of hydrogen-bonded networks that are stabilized by the inclusion of guest molecules within cavities left by the lattice. Because excellent compendia [28–30] are available that describe these water clathrates extensively, only a few conclusions, important in the context of this discussion, are mentioned. The versatility of water as a building unit is seen by the number of different clathrating structures observed to date. In each of these, the water molecules are completely hydrogen bonded with each oxygen surrounded by four nearest neighbor hydrogens. Thus in a transition from one clathrate structure to a second, the number of hydrogen bonds would not change; the size and geometry of the included guest molecule appear to be the important factors determining the form of the water lattice.

Apolar gases form clathrates with water at elevated gas pressures, although the decomposition pressures for many of the larger molecules are below 1 atm at 0°C (Table 4). Only two structures, designated type I and II, are commonly observed for the apolar gas hydrates; the greater structural variety is found when alkyl salts or polar compounds are the guest molecules [29, 30]. The type I hydrate contains cavities of two radii, 5.1 and 5.8 Å. With every cavity occupied, the limiting composition of the hydrate would

TABLE 4 PROPERTIES OF GAS HYDRATES[a]

Guest Molecule	Dissociation Pressure of Hydrate at 0°C	Decomposition Temperature at 1 atm
Type I hydrates		
Ar	105 atm	−42.8°C
CH_4	26	−29.0
Kr	14.5	−27.8
C_2H_6	5.2	−15.8
Xe	1.5	−3.4
C_2H_5F	0.70	+3.7
Cl_2	0.33	+9.6
Br_2	0.06	+20
Type II hydrates		
C_3H_8	1.0	—
C_2H_5Cl	0.26	—
CH_3CHCl_2	0.07	—

[a] These data were collected by R. M. Barrer [28].

be $X_4(H_2O)_{23}$. When only the larger cavities are filled, the stoichiometric formula becomes $X_3(H_2O)_{23}$. Cavities of two sizes, with radii of 5.0 and 6.7 Å, are also found in the type II structure. The compositions of the completely and partially filled structures are $X_3(H_2O)_{17}$ and $X(H_2O)_{17}$. The relatively large pressure needed to stabilize most gas hydrates is due to the requirement for occupation of a sufficiently large fraction of the ca ities within the lattice. This concentration effect is dramatically evident in the stabilization of type II hydrates by a second gas (Table 5) called the "help" gas [31]. This cosolute raises the decomposition temperature by occupying the smaller cavities into which the primary solute does not fit. It is incorrect, therefore, to assume that the water clathrates are inherently unstable. Many are more stable than ice I as indicated by their higher melting points (Tables 4 and 5). The stability possible with a water lattice is observed in a clathrate of tri-n-butylsulfonium fluoride [33], which melts at 5.6°C. Two $(C_4H_9)_3S^+$ ions are located within a single cavity of the water network.

TABLE 5 DECOMPOSITION TEMPERATURES OF TYPE II HYDRATES AT 1 atm PRESSURE[a]

Help Gas	CH_3COCH_3	CH_2Cl_2	$CHCl_3$	CCl_4
Ar	−8.0	−7.0	−4.8	−1.6
Kr	−5.0	+6.2	+9.0	+11.3
Xe	+3.0	+8.6	+10.9	+13.7

[a] Ref. 32.

The apolar side chains, however, extend into the aqueous matrix with the charged sulfur atoms confined to the center, separated from one another by the Van der Waals distance of 3.49 Å.

There are few thermodynamic data for the dissolution of apolar gases into preformed water clathrates. As detailed by Van der Waals and Plateeuw [34], much of the data that are available may be unreliable. A major experimental obstacle is the instability of the empty lattice. Thus the physical process that is most accessible to measurement is:

$$X + nH_2O \rightleftharpoons X(H_2O)_n \tag{38}$$

TABLE 6 ENTROPIES AND ENTHALPIES OF GAS VAPORIZATION FROM WATER CLATHRATES AT 0°C

Solute	ΔH_2^v(kcal/mole)[a]	ΔS_2^v(cal/mole deg)[a]
Type I hydrates		
H_2S	6.7	20
CH_4	5.6	21
C_2H_6	6.9	24
Cl_2	7.9	21
CH_3Br	9.5	24
Type II hydrates		
CH_3I	10.7	23
C_2H_5Cl	12.1	30
C_3H_8	9.7	26
$CHCl_2F$	11.9	28
$CBrClF_2$	11.6	28
Ar	7.1[b]	27[b]
Kr	6.7[b]	25[b]
Xe	6.9[b]	27[b]

[a] Calculated by Child from compiled data [36].
[b] Average values computed over the temperature range −78 to −10°C with CCl_4 present as help gas [37].

where gas solution and lattice formation are observed simultaneously. The process we wish to consider is:

$$X + (H_2O)_n \rightleftharpoons X(H_2O)_n \tag{39}$$

in which the gas dissolves into an empty, but otherwise constructed, lattice. A procedure for estimating the thermodynamic parameters of the process depicted in equation 39 from data obtained with the reaction of equation 38 has been proposed by Child [35]. The assumptions required for the calculation have no direct validation, and as a result the derived parameters are useful only for a qualitative discussion of possible trends. There are, fortunately, data available that were obtained directly with preformed hydrates. Barrer and Edge [37] determined the solubility of gases into the smaller cavity of a type II clathrate that had been previously stabilized with carbon tetrachloride. Although data were not collected above $-10°C$, the results (Table 6) are qualitatively similar to and so supportive of the values calculated by Child [36]. We now consider the results collected in Table 6 in terms of the theory of Frank and co-workers for liquid solutions.

Lattice stabilization, or an equilibrium shift towards structured solvent, cannot be postulated for the process of gas dissolution into a preformed clathrate, which is already fully structured. On this basis it could be argued that were a sizeable proportion of the entropy and enthalpy of gas vaporization from liquid water due to the collapse of an icelike structure, then the corresponding parameters obtained with crystalline hydrates should be significantly less positive. The appropriate values for this comparison are taken from Tables 1 and 6 and collected in Table 7. The two sets of data are not directly comparable, because the experimental temperatures were different. However, the solution of gases into the crystalline hydrate has an

TABLE 7 COMPARISON OF SOLUTE VAPORIZATION FROM LIQUID AND CLATHRATE WATER

Gas	ΔH_2^v(kcal/mole)		ΔS_2^v(e.u./mole)	
	Liquid[a]	Clathrate[b]	Liquid[a]	Clathrate[b]
Ar	2.8	7.1	30	27
Kr	3.6	6.7	33	25
Xe	4.4	6.9	34	27
CH_4	3.2	5.6	32	21
C_2H_6	4.4	6.9	35	24
C_3H_8	5.6	9.7	39	26

[a] Obtained from References 11 and 38; $t = 25°C$.
[b] Obtained from Table 6; $t = 0°C$ for hydrocarbons, $t = -10°C$ for inert gases.

apparently small heat capacity change [37] so that only a small correction would be required to bring the clathrate parameters to the "correct" temperature; that is, comparison as an approximation is valid. Contrary to the prediction just formulated, the enthalpies obtained with the clathrate solutions are not significantly more negative. Moreover, although the liquid entropies are more positive, the differences between liquid and clathrate is of the same magnitude as those between clathrate and carbon tetrachloride, for those cases where a comparison can be made (Table 1). We may conclude that between liquid and clathrate water there is not a large difference that can be clearly assigned to a structuring process. In addition, a comparison of carbon tetrachloride and water clathrate solutions reveals differences that cannot be attributed to structural breakdown, because this process presumably does not occur in either solvent. Thus the observation of greater positive entropies and enthalpies obtained with liquid water as compared to carbon tetrachloride cannot be utilized *per se* to argue for structural stabilization of water by apolar solutes. In a later section we return to this point and attempt to show that the observed differences can be explained without invoking any stabilization model.

2.3 Solution Partition and the Apolar Bond

As proposed by Kauzmann [1] and discussed by Klotz [6], the thermodynamic data presented in the previous sections can be used for the formulation of a new bonding situation in macromolecules, the apolar bond. We start with the three assumptions that apolar amino acid side chains can be represented as independent solutes, that the aqueous environment about a polypeptide chain is identical to bulk water, and that a protein interior, composed of the apolar portions of a macromolecule, is analogous to an apolar liquid solvent. Were these assumptions valid, the "transfer" of an apolar side chain from the aqueous environment into the macromolecular interior would closely resemble the transfer of an apolar solute from water into carbon tetrachloride or some other apolar liquid; the thermodynamic parameters describing this latter partition would apply to the macromolecular folding process as well. It is evident that these three assumptions cannot be correct; but the better they are as approximations, the stronger are any qualitative arguments based on them. This is the liquid-liquid partition model proposed by Kauzmann. Because the partition free energies for apolar solutes are negative (e.g., Table 3), Kauzmann reasoned that removing apolar side chains from the aqueous environment by tucking them into the protein interior should be a spontaneous process with the released free energy contributing to the stabilization of the polypeptide's three-dimensional structure. Using a negative free-energy change as the criterion of a bonding situation, this partition of the apolar side chains due to the non-covalent solute-solvent interactions would constitute a bond. In contrast to other accepted bonds, this one is endothermic (Table 1) with the entropy providing the driving force. An entropic bond offers, however, no conceptual difficulties.

In proposing the liquid-liquid partition model, Kauzmann adopted the water lattice stabilization theory of Frank and co-workers and suggested that collapse of icelike structure is a primary driving force of the apolar bond. This suggestion, which has acquired great importance in most textbooks, is probably incorrect. As discussed previously, the free energy of solute dissolution into water contains no contribution from the positional shift of a water species equilibrium. Kauzmann's model is, however, a thermodynamic formulation and, therefore, does not depend on the mechanism postulated for interpretation of the thermodynamic parameters observed. Regardless of the physical explanation, apolar molecules tend to escape from water into an apolar liquid. The validity of the liquid-liquid partition model depends only on how closely the assumptions required for its formulation correspond with physical reality, that is, on how similar the relocation of an amino acid side chain into a protein interior is to the partition of an apolar molecule between liquid phases. Currently, no quantitative arguments have been proposed for testing these hypotheses. It is, therefore, difficult to answer questions such as how applicable the thermodynamic parameters of Table 3 are to calculations of conformational stabilities, and whether one or another organic liquid [113] would serve as a better model of the protein interior.

A second formulation of the apolar bond, the liquid-clathrate partition model, was proposed by Klotz [6]. The formalism of this model is similar to that of Kauzmann's, with the important difference lying in the postulated structure of the water in contact with the protein surface. Klotz has suggested that this water does not resemble the bulk liquid, but rather may be structured in a manner similar to the water clathrates described in the previous section. On this basis, the partition of apolar residues would be best described by the transfer of apolar solutes between water clathrates and apolar liquids. Unfortunately, few reliable data on the solubility of apolar molecules in clathrate water are available, limiting the usefulness of this model. To this reviewer's knowledge the data collected in Table 6 constitute the total available information. Most of these values were obtained by a still-unsubstantiated extrapolation of experimental results and are for this reason not sufficiently reliable for free-energy calculations. The last three entries of Table 6 were, however, obtained by direct measurements with preformed water clathrates [37], and for argon, sufficient liquid solubility data are available to estimate liquid-clathrate partition free energies. Although the calculated results are not ambiguous (Table 8), general conclusions to be

TABLE 8 VAPORIZATION FREE ENERGIES OF
ARGON FROM CCl_4, WATER, AND TYPE II HYDRATE
AT $0°C$

Solvent	ΔG_2^v(kcal/mole)	ΔG^{tr}(kcal/mole)
$CCl_4{}^a$	−3.6	
H_2O^b	−5.4	+1.8 $CCl_4 \rightleftharpoons H_2O$
Type II hydrate^c	−0.3	−5.1 $H_2O \rightleftharpoons$ hydrate
		−3.3 $CCl_4 \rightleftharpoons$ hydrate

^{a} Reeves and Hildebrand [40].
^{b} Battino and Clever [39].
^{c} Barrer and Edge [37]. The hydrate utilized had been previously
stabilized with CCl_4. ΔG_2^v at $0°C$ was calculated on the assumption
that ΔH_2^v and ΔS_2^v were temperature independent. This appears to
be approximately true over the range −78 to −10°C.

derived from them must be held as tentative. The transfer of argon to a
preformed water clathrate from either liquid water or carbon tetrachloride
is a favorable process and is, moreover, exothermic, as can be seen by
comparison of results presented in Tables 1 and 7. Thus if the data for argon
can be applied generally to other solutes and if the protein surface can
stabilize a hydrate structure in the same manner as the help gas stabilizes a
type II clathrate, then the liquid-clathrate model of the apolar bond would
be thermodynamically sound. Once again, however, the applicability of
this model to a protein system depends on the validity of the assumptions
required for its formulation, and as in the case of the liquid-liquid partition
model, there are no quantitative arguments with which to test this
question.

Although the Klotz and Kauzmann models for the apolar bond differ in
only one aspect, the assumed nature of the water in contact with the protein
surface, the consequences of this difference are profound. The apolar bond
described by the liquid-liquid partition model is endothermic, and stabilization would be achieved on burying the apolar amino acid residues into the
protein interior. The bond described by the liquid-clathrate partition model
is exothermic, and stabilization comes from exposure of the apolar amino
acid residues to the aqueous environment. There are, therefore, two possible
criteria with which to compare and evaluate these models: The sign of the
enthalpies associated with protein denaturation or disaggregation, and the
location of apolar residues in native proteins. The complexity of protein
structural transitions, which involve a number of physical processes, preclude
direct utilization of thermodynamic data in evaluating the suitability of the
two models. An excellent start has been made in attempting to unravel the
many possible contributions to conformational stability [4, 13]. This analysis,

however, must be considered as still primitive and presently of problematic utility for isolating the parameters associated with apolar bonding. Thus the thermodynamic criterion cannot yet be considered as accessible.

Because of the achievements of x-ray crystallographers, topographical analysis of an ever increasing number of proteins is now feasible. Nonetheless, application of the second criterion for comparison of the two apolar bond models, the location of apolar amino acids, yields ambiguous results. It has been generally assumed that an overwhelming majority of apolar residues are not in contact with the aqueous environment [41]. On the basis of this belief, the Kauzmann model has received the greatest support. It is now recognized, however, that although protein interiors are overwhelmingly apolar, apolar regions are to be found on the protein surface as well [114], and it has been argued that substantial fractions of the surfaces on four proteins are derived from apolar amino acids [42]. There are, clearly, problems in any attempt to assign internal versus external locations, that is, what defines the protein interior and how to classify partly buried residues. Lee and Richards [115] have proposed a semiquantitative approach to the description of surface positions that obviates these problems in part. Using atomic coordinates supplied by crystallographic data and assigning hard sphere volumes to atoms, they constructed and stored into computer memory space-filling models for a number of proteins. They then considered whether individual atoms of the protein were accessible to water by moving a sphere with the approximate collision radius of a water molecule over the space-filling model. A point on the surface of an atom attached to the protein was deemed accessible to solvent if the water sphere could touch that point without simultaneously overlapping the volume boundary of any other protein atom. Replacing integration with an empirical summing technique, the fraction of the individual atom's surface area accessible to the solvent was estimated. With these fractions for all of the protein's atoms, the acessible protein surface area was computed, together with the contributions of the various atoms to that surface. Then, defining carbon and sulfur atoms as apolar and nitrogen and oxygen as polar, the apolar contributions to the surface areas of ribonuclease-S, lysozyme, and myoglobin were estimated as 46%, 41%, and 52%, respectively. Although these estimates should be considered cautiously, they suggest that an appreciable fraction of the protein surface is apolar.

Regardless of its accuracy, however, no procedure for classifying internal versus external positions of apolar residues can yield results directly applicable for a judgement on the relative merits of the two available partition models. Both apolar bonds are formulated in the macroscopic language of thermodynamics, and as yet there is no microscopic description of this bond that is comparable to those available for other bonds. There is, therefore, the dilemma of extrapolating from the physical picture of apolar residue groupings to the intellectual construction of energetic stabilization. For example, the striking regularity observed in protein structures is the almost complete absence of charged and polar side chains within the interior [41, 114]. This distribution may be due to the energy gained by removing apolar side chains from water, as suggested by the liquid-liquid partition model. On the other hand, it may be argued that the energy gained from placing charged and polar residues at the interface for maximal water interaction is the driving force that results, by a subtractive process, in an apolar interior. It is, therefore, logically difficult to assume that the observation of an apolar protein interior supports the liquid-liquid partition model. Similarly, the apolar residues observed on the protein surface could be taken, on the basis of the liquid-clathrate model, as a stabilizing configuration; but it is also possible that these surface regions arise from the geometric limitations imposed by the folding process and need not contribute any stability. Lee and Richards [115] have proposed an argument that may lead to an alleviation of this logics problem. In addition to native proteins, they computed accessibilities for the atoms of hypothetical, extended polypeptide chains. Comparing the two sets of values, they found a greater decrease in the accessibility of the apolar atoms as compared with the polar atoms for the change extended to folded conformation. The differences observed were not, however, great enough to suggest that either an exterior or an interior position for an apolar residue would be energetically preferred, and their calculations at the present level of sophistication would not constitute support for one of the apolar bond models over the other.

The arguments presented so far have been largely qualitative, and, thus, no definitive conclusions regarding the correct formulation of the apolar bond has been possible. Determining the adequacy of a model described in terms of any solution process clearly requires as a minimum a quantitative understanding of solutions. We shall, therefore, postpone further discussion of the apolar bond and return to the problem of liquid solutions and the quantitative evaluation of their properties.

3 SCALED PARTICLE THEORY AND APOLAR GAS SOLUTIONS

In order to derive rigorously the thermodynamic parameters that characterize a dilute binary solution, the potential functions describing solute-solvent and solvent-solvent interactions must be first determined. With the correct functions in hand, entropies and enthalpies may be calculated by the methods of statistical thermodynamics. Both steps of this derivation are so formidable, however, that every current theory is based on various assumptions and simplifications, and none can be considered better than semi-empirical. This raises the problem of what criteria are useful for determining the success of any one theory. Ability to mimic experimental data is not sufficient, since few reasonable models containing adjustable parameters cannot be adequately tailored. On the other hand, a model based on sound premises may yield expressions that correlate poorly with experiment because of the introduction of computational simplifications required to perform the necessary mathematical manipulations. Nor is the degree of molecular detail introduced in the formulation of a theory a solely adequate measure of success, especially if there is no experimental verification of that detail. Two further criteria are certainly required; the further insight or intellectual sophistication gained from the derivation of the theory, and the ability of the proposed model to describe phenomena related to those directly under consideration, but not explicitly considered in the derivation.

For example, the theory of Némethy and Scheraga [43, 44], the most ambitious attempt yet made at a description of apolar molecules in aqueous solution, is successful when viewed with the first two of these criteria. The final Gibbs free energies and the enthalpies calculated by these authors agree reasonably with experimental data, and the wealth of theoretical detail is impressive. There are, nonetheless, unsatisfactory aspects to this theory. Although the molecular structure of water is poorly understood [19, 20], a detailed description of water was required in the approach taken by Némethy and Scheraga. As a result, at one point in the derivation, physical constructions of "water clusters" were built for use in "hand counting" hydrogen bonds. Other required parameters, not available from experimental results, were introduced after crude estimation; finally, five theoretical parameters were left as adjustable. Even so, the developed theory was successfully extended to further properties of water or water solutions only by the introduction of additional adjustable parameters. These deficiencies arose undoubtedly from the sophistication with which the problem was attacked. In the future when detailed information concerning the structure of liquid water is available, the start made by Némethy and Scheraga should prove invaluable. In the meantime, theories that do not require such an extensive informational input may be more useful. We, therefore, next consider an empirical approach of great simplicity, which adequately reproduces a number of experimental observations.

3.1   The Scaled Particle Theory of Fluids

The scaled particle theory of dense fluids describes collections of spheres that have finite volumes and intermolecular potentials of zero at every point of separation except contact, where the potential rises abruptly to infinity; that is, the theory describes hard sphere liquids. The properties of this idealized liquid are then applied to real systems by the introduction of some finite intermolecular potential, such as the square well or the Lennard-Jones 6-12 functions. Real liquids are, thus, treated as Van der Waals fluids, an approximation that had fallen into disuse, but which has received renewed attention in the past decade [45]. Although the limitations imposed by the hard sphere approach would appear to be severe, a surprisingly large number of liquid properties can be reproduced adequately. Moreover, in the Van der Waals approximation the entropy of fluid phase changes depends primarily on the molecular size. Hence the disorder of those real liquids that behave as Van der Waals fluids is relatively independent of the intermolecular potentials once the liquid volume has been established. This is of theoretical advantage because molecular sizes are more easily estimated than are the intermolecular attractions.

The details of the scaled particle theory have been thoroughly described by Reiss [46], thus only a cursory derivation is presented here. We assume a pure liquid of hard spheres with the diameter δ_1, and ask how much work is required to place an additional sphere of any diameter δ into the liquid interior. Because the spheres have no attractive interaction, the only work in the solution process is that required to open a sufficiently large cavity in the solvent. This hole may be defined as the excluded cavity, a region in the fluid that contains no solvent centers; the radius of the excluded cavity is $(\delta + \delta_1)/2$ (Fig. 2), and its volume is $\pi(\delta + \delta_1)^3/6$. We initially consider the physically impossible sphere with a diameter $\delta \leqslant 0$. For this case the

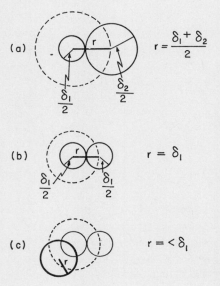

(a) $\quad r = \dfrac{\delta_1 + \delta_2}{2}$

(b) $\quad r = \delta_1$

(c) $\quad r = < \delta_1$

Figure 2 Excluded cavities for three arguments of the excluded radius.

excluded cavity has a radius $r \leqslant \delta_1/2$. Because this is smaller than the radius of any solvent sphere, the required volume element can contain no more than one spherical center (Fig. 2), and the probabilities of finding none or one center in this volume element are related by:

$$p_0(r) + p_1(r) = 1 \qquad r \leqslant \frac{\delta_1}{2}. \tag{40}$$

The probability of finding any solvent center in the excluded cavity is:

$$p_1(r) = \frac{\pi}{6} (\delta + \delta_1)^3 \rho_1 \qquad r \leqslant \frac{\delta_1}{2} \tag{41}$$

where the solvent density ρ_1 is defined by:

$$\rho_1 = \lim_{N_1, V \to \infty} \frac{N_1}{V} \tag{42}$$

with N_1 the number of spheres in the total liquid volume V. The probability of finding no centers in the excluded cavity is, therefore:

$$p_0(r) = 1 - \frac{\pi}{6} (\delta + \delta_1)^3 \rho_1 \qquad r \leqslant \frac{\delta_1}{2}. \tag{43}$$

The work required to open this cavity may be obtained by invoking Boltzmann statistics:

$$W' = -kT \ln \left[1 - \frac{\pi}{6} (\delta + \delta_1)^3 \rho_1 \right] \qquad r \leqslant \frac{\delta_1}{2} \tag{44}$$

since:

$$p_0(r) = e^{-W'/kT}. \tag{45}$$

Because the liquid volume change during the solution process is of negligible significance, W' is a good approximation of the Gibbs free energy.

Although equation 45 is exact, it is not applicable to physically significant systems for which δ must be greater than zero. When, however, $r > \delta_1/2$, the excluded volume element can contain more than one fluid center; equation 40 is no longer valid and must be replaced by:

$$\sum p_i(r) = 1. \tag{40'}$$

Because a closed expression for $p_i(r)$ is not readily available, an exact relationship corresponding to equation 43 has not been derived. Instead, it is assumed that the work required to open an excluded cavity of any radius r is approximated by the power series:

$$W(r) = C_0 + C_1 r + C_2 r^2 + C_3 r^3 \tag{46}$$

where C_0 through C_3 are arbitrary constants that must be evaluated. (After completion of the calculations to be described subsequently, Tully-Smith and Reiss [116] refined the theory by extending this power series. The less-precise relationship of equation 46 is, however, sufficient for our purposes, because the introduction of additional parameters in equation 46 would improve the results only slightly.) If this approximation is valid, equation 46 should hold for the point $r = \delta_1/2$ where it is equivalent to equation 44, and also for macroscopic values of r where it is known from thermodynamics that:

$$W'' = \tfrac{4}{3}\pi r^3 p + 4\pi r^2 \tau \frac{1 - k_1}{r} + k_0 \tag{47}$$

in which p and τ are the liquid pressure and surface tension, and k_0 and k_1 are constants with no significance in the context of this discussion.

The coefficients of equation 46, C_0 through C_3, can be evaluated by utilization of equations 44 and 47. Comparing 46 with 47 suggests the identity:

$$C_3 = \frac{4\pi p}{3}. \tag{48}$$

Since it is also true that [46]:

$$\left. \begin{aligned} \frac{\partial W(r)}{\partial r} &= \frac{\partial W'}{\partial r} \\ \frac{\partial^2 W(r)}{\partial r^2} &= \frac{\partial^2 W'}{\partial r^2} \end{aligned} \right\} \quad r = \frac{\delta_1}{2} \tag{49}$$

combining 46 with 44, 47, 48, and 49 yields:

$$W(r) = kT\left[3\left(\frac{\xi}{1 - \xi}\right)\left(\frac{\delta^2}{\delta_1^2} + \frac{\delta}{\delta_1}\right) + \frac{9}{2}\left(\frac{\xi}{1 - \xi}\right)^2 \frac{\delta^2}{\delta_1^2} - \ln(1 - \xi) \right] + \frac{\pi p \delta^3}{6} \tag{50}$$

where ξ, the fraction of the liquid volume actually covered by the solvent sphere, is:

$$\xi = \frac{\pi \delta_1^3 \rho_1}{6}. \tag{51}$$

The equation of state for a hard sphere liquid is more useful for our purposes and is obtained as follows. Assume that the volume element of radius r contains no solvent centers. The spherical shell of thickness dr that surrounds the empty cavity has a volume of $4\pi r^2 \, dr$. The probability that this shell contains one or more solvent centers is given by:

$$4\pi r^2 \rho_1 g(r) \, dr \tag{52}$$

where $g(r)$ is a liquid distribution function so defined that $\rho_1 g(r)$ is the density of solvent centers in contact with the excluded cavity of radius r. The probability, $p_0(dr)$, that the spherical boundary shell is empty becomes:

$$p_0(dr) = 1 - 4\pi r^2 \rho_1 g(r)\, dr \tag{53}$$

and the probability, $p_0(r + dr)$, that both the excluded cavity and the surrounding shell are simultaneously empty is the product of the individual probabilities:

$$p_0(r + dr) = p_0(r)[1 - 4\pi r^2 \rho_1 g(r)\, dr] = p_0(r) + \frac{\partial p_0(r)}{\partial r}\, dr. \tag{54}$$

On rearrangement and integration equation 54 becomes:

$$p_0(r) = \exp\left(-\int_0^r 4\pi r^2 \rho_1 g(r)\, dr\right). \tag{55}$$

Invoking Boltzmann statistics:

$$W(r) = 4\pi kT \int_0^r r^2 \rho_1 g(r)\, dr \tag{56}$$

$$\frac{\partial W(r)}{\partial r} = 4\pi kT r^2 \rho_1 g(r). \tag{57}$$

According to the virial theorem, the equation of state for a hard sphere fluid mixture is [47]:

$$\frac{p}{kT} = \sum \rho_i + \frac{2}{3}\pi \sum \rho_i \rho_j r_{ij}{}^3 g(r_{ij}); \qquad r_{ij} = \frac{\delta_i + \delta_j}{2}. \tag{58}$$

When $\delta_i = \delta_j$, $g(r_{ij})$ becomes the distribution function $g(r)$ of the pure liquid, and the equation of state reduces to:

$$\frac{p}{kT} = \rho_1 + \frac{\pi}{6}\rho_1{}^2 \delta_1{}^3 g(r). \tag{59}$$

If the radius r of the excluded cavity is set equal to δ_1, equations 50, 57, and 59 may be combined to obtain the equation of state for a pure hard sphere fluid:

$$\frac{p}{kT} = \frac{1 + \xi + \xi^2}{(1 - \xi)^3}\, \rho_1. \tag{60}$$

An additional relationship that may be obtained directly from equation 60 is the hard sphere fluid free volume.

$$v^f = \frac{(1 - \xi)^3}{1 + \xi + \xi^2}\, V_M \tag{61}$$

where V_M represents the molar volume.

An equation of state for mixtures of hard spheres may be obtained by a similar procedure. Thus the probability of finding a fluid center in an excluded cavity with a radius smaller than the smallest fluid particle of diameter δ_1 is:

$$p_1(r) = \frac{\pi}{6} (\delta + \delta_1)^3 \rho \qquad r < \frac{\delta_1}{2} \qquad (41')$$

where the density ρ is now defined by:

$$\rho = \lim_{V, N_1, N_2 \ldots \to \infty} \frac{N_1 + N_2 + \cdots N_n}{V}. \qquad (42')$$

Applying the same arguments of the previous paragraphs the equation of state can be shown to be [47]:

$$\frac{p}{kT} = \frac{6}{\pi} \left[\frac{\theta_0}{1 - \theta_3} + \frac{3\theta_1\theta_2}{(1 - \theta_3)^2} + \frac{3\theta_2^2}{(1 - \theta_3)^3} \right] \qquad (62)$$

with the θ parameters defined as:

$$\theta_m = (\pi/6) \sum_i \delta_i{}^m \rho_i; \qquad m = 0, 1, 2, 3. \qquad (63)$$

The distribution function for a binary hard sphere fluid, which will be required in a later argument, may also be obtained [47]:

$$g_{12}(r) = \frac{1}{1 - \theta_3} + \frac{3\theta_2}{(1 - \theta_3)^2} \left(\frac{\delta_1\delta_2}{\delta_1 + \delta_2} \right) + \frac{3\theta_2^2}{(1 - \theta_2)^3} \left(\frac{\delta_1\delta_2}{\delta_1 + \delta_2} \right)^2. \qquad (64)$$

The equations of state for hard sphere fluids (eqs. 60 and 62) may be applied to those real liquids for which the Van der Waals approximation is valid [46]. For this purpose the hard sphere diameters must be correlated with the molecular size, and the number densities assigned. In the Van der Waals approximation the intermolecular potential of real molecules is separated into two terms: a repulsive contribution due to a hard core center and a soft term arising from electrostatic attraction. Equations of state for gases are treated with a similar convention in which the hard core term is specified by the molecular collision diameter, the distance between the molecular center and the unique point of zero potential (Fig. 3). The hard sphere diameters δ_i of equations 51 and 63 will, therefore, be replaced by the collision diameter σ, which can be estimated by standard procedures from the virial coefficients of gases [49].

Once the minimum permissible or closest packed volume of a sphere collection is specified by the appropriate collision diameters, the fluid density may be treated as a function of the intermolecular attractions; that is, the volume of the collection is determined by the soft contribution

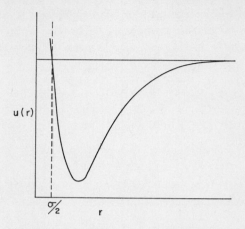

Figure 3 Definition of the collision diameter.

to the intermolecular potential. The precise analytic form this function should take is not, however, known. Rather than using one of the various arbitrary assignments for this function, the hard sphere density in the derived equations of state is replaced empirically with experimental liquid densities. By this procedure the intermolecular attractive forces are assigned effectively, though not explicitly. Susbtituting σ and the experimental volume V into equations 51 and 63 yields the new definitions:

$$\xi = \frac{\pi}{6V_M} \sigma^3 N_M \tag{51'}$$

$$\theta_m = \frac{\pi}{6V} \sum_i \sigma_i{}^m N_i; \qquad m = 0, 1, 2, 3 \tag{63'}$$

where N_M and V_M are Avogadro's number and the molar volume of the pure liquid and N_i is the number of ith species molecules in V, the observed mixture volume.

Equation 60, together with the definition introduced in equation 51', has been used to derive expressions for a number of pure liquid properties. Details of the derivations and of the correlation between theory and observation have been collected by Reiss [46], and for this reason are not reconsidered here. Only those expressions and procedures that are of use for a discussion of liquid solutions are described. The isothermal compressibility of a pure liquid is defined as:

$$\beta = -\frac{1}{V_M}\left(\frac{\partial V_M}{\partial \rho}\right)_T. \tag{65}$$

Differentiation of equation 60 and substitution of the result into equation 65 yield directly:

$$\beta = \frac{V_M}{RT} \frac{(1 - \xi)^4}{(1 + 2\xi)^2}.$$ (66)

This equation reproduces the isothermal compressibility of organic and elemental liquids and fused salts remarkably well, but begins to break down for liquids in which hydrogen bonding is important [46]. The isobaric expansivity:

$$\alpha = \frac{1}{V_M}\left(\frac{\partial V_M}{\partial T}\right)_p$$ (67)

may also be derived directly from equation 60:

$$\alpha = \frac{1 - \xi^3}{(1 + 2\xi)^2} \cdot \frac{1}{T}.$$ (68)

As mentioned previously, the entropy of phase transitions in a Van der Waals fluid may be taken to be independent of the soft, attractive interaction. Insofar as real liquids behave as Van der Waals fluids, the scaled particle theory should be useful in calculations of entropy changes. This was shown to be true by Yosim and Owens [50], who calculated the vaporization heats of pure liquids with the following hypothetical entropy cycle. The vaporization of a liquid is arbitrarily divided into three steps.

1. The attractive potential between the molecules of the liquid is discharged leaving a hard sphere fluid that occupies the same volume as the real liquid. The isochoric entropy change for this process is $S_l^h - S_l$ where the superscript h refers to the hypothetical hard sphere fluid.

2. The hard sphere liquid is then expanded isothermally at the boiling point T_b, where the pressure is 1 atm. Because the hard sphere gas is in equilibrium with the liquid:

$$dH^h - T_b\, dS^h = 0.$$ (69)

Substitution of the internal energy U into equation 69 and rearrangement yield:

$$\left(\frac{\partial S^h}{\partial V}\right)_{T_b} = \frac{1}{T_b}\left[\left(\frac{\partial U^h}{\partial V}\right)_{T_b} + p\right].$$ (70)

However, for a hard sphere system with no intermolecular attractions the internal energy does not depend on the volume, that is, $(\partial U^h/\partial V)_{T_b} = 0$. Equation 70 may then be integrated to obtain:

$$\Delta S^h = \frac{1}{T_b}\int_{V_l}^{V_g} p\, dV.$$ (71)

3. In the final step the attractive forces between the molecules in the gas state are recharged. The isochoric entropy is $S_g - S_g{}^h$.

The net entropy change for the overall process is the sum of the three entropy differences:

$$\Delta S^v = (S_l{}^h - S_l) + (S_g - S_g{}^h) + \frac{1}{T_b} \int_{V_l}^{V_g} p \, dV. \tag{72}$$

If the intermolecular attraction in the liquid is weak and the potential function centrosymmetric, then discharging the attractive interaction should result in only minor configurational rearrangements. Thus, entropy change because of configurational relaxation should be negligible. All other possible entropy contributions to the charging and discharging processes would be due to electronic changes within the molecules and would be identical in both the liquid and gas phases. We, therefore, make the approximation:

$$(S_l{}^h - S_l) + (S_g - S_g{}^h) \approx 0 \tag{73}$$

so that equation 72 reduces to:

$$\Delta S^v \approx \frac{1}{T_b} \int_{V_l}^{V_g} p \, dV. \tag{74}$$

The integral in equation 74 refers to the pure hard sphere fluid only and can be evaluated with the use of equation 60.

$$\frac{\Delta S^v}{R} = \ln \left[\frac{(V_g/V_l) - \xi}{1 - \xi} \right] - \frac{3}{2} \xi \left\{ \frac{(2V_g/V_l) - \xi}{[(V_g/V_l) - \xi]^2} - \frac{2 - \xi}{(1 - \xi)^2} \right\} = \frac{\Delta H^v}{RT_b}. \tag{75}$$

Since $(V_g/V_l) \gg 1 > \xi$, this relationship reduces to:

$$\frac{\Delta H^v}{RT_b} = \ln \frac{V_g}{V_l(1 - \xi)} + 3 \left(\frac{\xi}{1 - \xi} \right) + \frac{3}{2} \left(\frac{\xi}{1 - \xi} \right)^2. \tag{76}$$

Assuming ideal gas behavior to calculate V_g, assigning V_l on the basis of measured liquid densities and utilizing collision diameters obtained from gas virial coefficients, Yosim and Owens calculated the vaporization heats collected in Table 9. The agreement between observed and calculated results is remarkably good, indicating that many liquids may be treated with the Van der Waals approximation. The correlation between experimental and theoretical values begins to break down at higher molecular weights and for liquids in which hydrogen bonding becomes important. It would be expected that the approximation of equation 73 does not hold in these cases.

TABLE 9 LIQUID HEATS OF VAPORIZATION[a]

Liquid	ΔH^v (cals/mole)		
	Calculated	Observed	Calculated/observed
He	26.8	19.4	1.38
Ne	427	414	1.03
Ar	1611	1558	1.03
Kr	2188	2158	1.01
Xe	3264	3020	1.08
H_2	227.5	215.8	1.05
N_2	1434	1333	1.08
O_2	1749	1628	1.07
F_2	1581	1562	1.01
Cl_2	5154	4878	1.06
Br_2	6869	7170	0.96
I_2	12673	9970	1.27
NO	3139	3293	0.95
CO	1409	1444	0.98
HCl	3363	3860	0.87
HI	5104	2724	1.08
CS_2	6324	6400	0.99
N_2O	3990	3958	1.01
SO_2	7086	5960	1.19
CH_4	2098	1988	1.06
C_2H_4	3499	3237	1.08
C_2H_6	2987	3157	0.95
C_4H_{10}	4676	5352	0.87
C_6H_6	7312	7353	0.99
C_6H_{12}	10430	7190	1.45
CH_3Cl	3634	5150	0.71
$CHCl_3$	9539	7020	1.36
CCl_4	10685	7170	1.49
CH_3OH	5923	8430	0.70
C_2H_5OH	7148	9220	0.78

[a] Results taken from reference 50.

Although full documentation has not been presented here, scaled particle theory can be successfully utilized for predicting a number of pure liquid properties [46]. This success is encouraging for the subsequent discussion of liquid solutions. We first, however, digress to consider the systematic assignment of collision diameters to molecules. Results such as those of Table 9 indicate that molecular radii assigned on the basis of gas data can be applied to liquids as well. Although this conclusion is satisfying, it would be, nonetheless, more consistent to base collision diameters on the properties of the liquids themselves, because the molecular proximities differ greatly in the two states. There is the additional practical advantage that more experimental data are available for liquid systems. To assign molecular size from the properties of liquids is easily done by solving one or more of the derived scaled particle theory equations for the collision diameter and then inserting the pertinent experimental result. By using the heat of vaporization data (Table 9), diameters can be calculated with equation 76. Although this equation cannot be explicitly solved, σ can be extracted indirectly by the Newton-Rafson procedure [51]. Other properties of pure liquids can be similarly employed, and a collection of calculated collisional diameters are presented in Table 10. These parameters are used in all subsequent discussion.

TABLE 10   COLLISIONAL DIAMETERS $\sigma(\overset{\circ}{A})$

Species	β^a	$\Delta H\,vap^b$	Other
He			2.56^c
Ne		2.71	
H_2		2.73	
H_2O			2.75^d
Ar	3.20		
O_2	3.33		
N_2	3.41		
NO		3.50	
Kr		3.58	
CO		3.63	
CH_4		3.72	
N_2O		3.81	
Xe		3.84	
C_2H_6		4.08	
SO_2		4.11	
C_2H_4		4.12	
CH_3Cl		4.23	
CS_2	4.46		
I_2		4.69	
CF_4			4.70^d
$n\text{-}C_3H_8$			4.72^e
$CHCl_3$	4.78		
C_6H_6	5.01		
CCl_4	5.14		
$n\text{-}C_4H_{10}$		5.24	
C_6H_{12}	5.33		

^{a} Obtained from liquid compressibility and reported in Reiss [46].
^{b} Obtained from heat of liquid vaporization and equation 76.
^{c} Obtained from analysis of gas properties [49].
^{d} Reported by Pierotti [16].
^{e} Calculated from liquid surface tension by the method of Mayer [52].

3.2 Application of Scaled Particle Theory to Binary Liquid Mixtures

We are now in a position to calculate the partial molar volumes, and vaporization entropies of apolar gases in liquid solution with equation 62, introducing no additional assumptions. Enthalpies can also be estimated, but as shall be evident, the derivation of this parameter requires the use of additional, somewhat unsatisfactory assumptions.

The partial molar volume of a solute is defined by:

$$\bar{V}_2 = \lim_{n_2 \to 0} \left(\frac{\partial V}{\partial n_2}\right)_{T, p, n_1, n_2, \ldots} \tag{77}$$

and can be obtained directly from equation 62. For a binary solution the index i in equation 63′ is set to 2. Equation 62 is then differentiated with respect to n_2, rearranged in order to extract $(\partial V / \partial n_2)$, and the proper limit taken to yield [88]:

$$\bar{V}_2 = \frac{\begin{aligned}1 + (r_e^3 + 3r_e^2 + 3r_e)\xi/(1-\xi) + (6r_e^3 + 9r_e^2) \\ \times [\xi/(1-\xi)]^2 + 9r_e^3[\xi(/1-\xi)]^3\end{aligned}}{1 + 7\xi/(1-\xi) + 15[\xi/(1-\xi)]^2 + 9[\xi/(1-\xi)]^3} V_i(1) \tag{78}$$

with r_e defined as the relative solute radius σ_2/σ_1 and $V_i(1)$ the molar volume of the solvent. Equation 78 was evaluated for solutions of apolar gases in water and carbon tetrachloride. The solvent volumes were determined from experimental densities [53, 54], and the collision diameters taken from Table 10. The calculated results are presented as the solid lines in Figure 4; experimental partial molar volumes were collected from the literature [88] and are plotted as points. It is seen that equation 78 reproduces the observed values in all cases, within experimental error, for both aqueous and nonaqueous solutions, with the possible exception of H_2 in CCl_4.

Pierotti [55] has calculated partial molar entropies of vaporization, beginning with equation 50. The derivation presented here, however, is based on equation 63 and the entropy cycle of Yosim and Owens described in the previous section [50, 56]. Solute vaporization is divided into the following arbitrary steps:

1. The attractive potentials of a binary liquid solution are discharged to yield a hard sphere mixture that occupies the volume of the real liquid. The isochoric entropy change is given by $S_i^h(1+2) - S_i(1+2)$, with h previously defined, and the designation $1 + 2$ referring to the mixture of solvent and solute.

2. The hard sphere solution is then expanded isothermally to form the gaseous binary mixture. The vaporization entropy is obtained with the

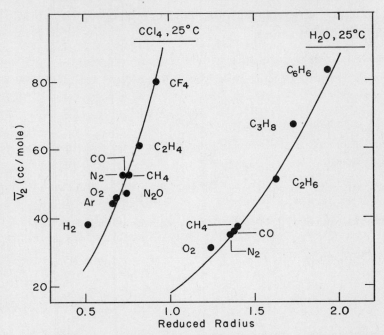

Figure 4   Partial molar volumes of apolar gases in water and carbon tetrachloride. The solid lines were calculated from equation 78; the points are experimental observations. This figure was taken from reference 88.

same argument used to reach equation 71 and is given by:

$$\Delta S^h(1+2) = \frac{1}{T} \int_{V_l(1+2)}^{V_g} p(1+2) \, dV. \tag{71'}$$

3. The hard sphere gaseous mixture is resolved into two pure phases requiring an entropy change due to the ideal mixing, $-\Delta S^M$.

4. The now pure hard sphere solvent is recondensed to a liquid with a volume corresponding to the volume of the real solvent, but the solute is left as a gas.

$$\Delta S^h(1) = \frac{1}{T} \int_{V_g}^{V_1(1)} p(1) \, dV. \tag{71''}$$

5. In the final step the attractive forces between the molecules of pure solvent liquid and pure solute gas are recharged. The isochoric entropy change is $[S_l(1) - S_l^h(1)] + [S_g(2) - S_g^h(2)]$.

In the overall process the solute is vaporized from the liquid solution with a net entropy change of:

$$\Delta S_2^v = [S_l(1) - S_l^h(1)] + [S_g(2) - S_g^h(2)]$$
$$+ [S_l^h(1+2) - S_l(1+2)] - \Delta S^M$$
$$+ \frac{1}{T}\left[\int_{V_l(1+2)}^{V_g} p(1+2)\, dV + \int_V^{V_l(1)} p(1)\, dV\right]. \quad (79)$$

We assume, as before, that the sum of the charging and discharging entropies is negligibly small:

$$\Delta S_2^v \approx -\Delta S^M + \frac{1}{T}\left[\int_{V_l(1+2)}^{V_g} p(1+2)\, dV + \int_{V_g(1+2)}^{V_l(1)} p(1)\, dV\right]. \quad (80)$$

Substitution of equations 3, 60, and 62 into equation 80 yields:

$$\Delta S_2^v = k\left[(\textstyle\sum N_i)\ln\left(\frac{V_g/V_l}{1-\theta_3}\right) + 3(\textstyle\sum N_i\sigma_i)\frac{\theta_2}{1-\theta_3} + \frac{3}{2}(\textstyle\sum N_i\sigma_i^2)\left(\frac{\theta_2}{1-\theta_3}\right)\right]$$
$$- kN_1\left[\ln\frac{V_g/V_1}{1-\xi} + 3\frac{\xi}{1-\xi} + \frac{3}{2}\left(\frac{\xi}{1-\xi}\right)^2\right] - k\textstyle\sum N_i\ln X_i \quad (81)$$

where ξ refers to the pure solvent liquid and θ to the solution (eqs. 51' and 63').

The partial molar entropy of solute vaporization is defined by:

$$\Delta \bar{S}_2^v = \lim_{n_2\to 0}\left(\frac{\partial \Delta S_2^v}{\partial n_2}\right)_{T,p,n_1}. \quad (82)$$

By performing the operations indicated in equation 82 and setting the standard state according to the rational thermodynamic scale, equation 81 is converted to:

$$\frac{\Delta \bar{S}_2^v}{R} = 1 + \ln\left[\frac{V_g}{V_1(1-\xi)}\right] + \left[\left(\frac{\xi}{1-\xi}\right)\left(\frac{\sigma_2}{\sigma_1}\right)^3 - \frac{\bar{V}_2/V_1}{1-\xi}\right]$$
$$\times \left[1 + 3\frac{\xi}{1-\xi} + 3\left(\frac{\xi}{1-\xi}\right)^2\right] + 3\left(\frac{\sigma_2}{\sigma_1}\right)\left(\frac{\xi}{1-\xi}\right)$$
$$\times \left[1 + \frac{\sigma_2}{\sigma_1} + \frac{3}{2}\left(\frac{\sigma_2}{\sigma_1}\right)\left(\frac{\xi}{1-\xi}\right)\right]. \quad (83)$$

As would be expected, equation 82 reduces to 76 when σ_2 is set equal to σ_1.

Because apolar gas solutions are dilute, the experimentally measured vaporization entropy ΔS_2^v closely approximates the partial molar vaporization entropy $\Delta \bar{S}_2^v$; thus experimental data may be directly compared with

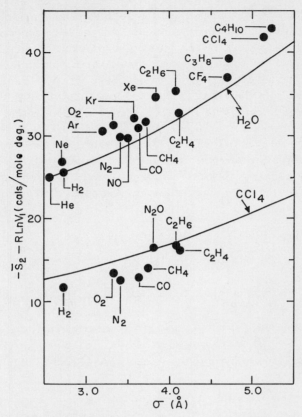

Figure 5   Vaporization entropies of apolar gases in water and carbon tetrachloride. The solid lines were calculated from equation 83. The experimental values shown as the points were obtained from references 11, 16, and 105.

the calculations based on equation 83. Results collected from the literature [11, 16, 105] are plotted in Figure 5 using the collision diameters assigned in Table 10. The calculated results are presented as solid lines in the same figure. As can be seen, equation 83 does predict the experimental results to a first approximation. The fit between theory and experiment is not exact, however. It should be realized that the original equation of state on which the derivation is based (eq. 62) is itself not exact, and that the approximation introduced in equation 80 results in an error of unknown magnitude. In spite of these uncertainties, the major difference between carbon tetrachloride and water solutions—the much greater solute vaporization entropies from water—is adequately produced by equation 83.

The final thermodynamic parameter to be considered is the solute vaporization enthalpy. Unlike the vaporization entropy, the enthalpy cannot be directly calculated on the basis of a hard sphere fluid model, because intermolecular attraction between solute and solvent is a primary factor in establishing the magnitude of the term. The problem of intermolecular attraction was circumvented for pure liquids in equilibrium with their vapors [50] by the use of equation 69, which allows the heat of vaporization to be calculated from the entropy. Although equation 69 must also apply to a solute in equilibrium with its vapor, the solute vaporization enthalpy $\Delta H_2{}^v$ contains the solute saturation concentration (eq. 81). For this reason, $\Delta H_2{}^v$ cannot be determined without prior knowledge of the vaporization Gibbs free energy. An alternate procedure for obtaining $\Delta H_2{}^v$ has been proposed by Boublik and Benson [57]. The internal energy change, ΔU, is considered over the same vaporization cycle used to derive equation 79. As a result:

$$\Delta U_2{}^v = [U_l(1) - U_l{}^h(1)] + [U_g(2) - U_g{}^h(2)]$$
$$+ [U_l{}^h(1+2) - U_l(1+2)] - \Delta U^M$$
$$+ \int_{V_l(1+2)}^{V_g} \frac{\partial}{\partial V} [U^h(1+2)] \, dV + \int_{V_g}^{V_l(1)} \frac{\partial}{\partial V} [U^h(1)] \, dV. \quad (84)$$

For a hard sphere fluid $(\partial U/\partial V)_T$ and ΔU^M are both zero:

$$\int_{V_l}^{V_g} \frac{\partial}{\partial V} [U^h(1+2)] \, dV + \int_{V_g}^{V_l(1)} \frac{\partial}{\partial V} [U^h(1)] - \Delta U^M = 0. \quad (85)$$

Moreover, for most gases the approximation of ideal behavior introduces only small errors so that $U_g(2) - U_g{}^h(2)$ may be ignored. Equation 84 then reduces to:

$$\Delta U_2{}^v = [U_l(1) - U_l{}^h(1)] + [U_l{}^h(1+2) - U_l(1+2)]. \quad (86)$$

Because the mechanical work of charging and discharging the intermolecular attractions in the solution and pure solvent is negligible, the internal energy and enthalpy are comparable and:

$$\Delta H_2{}^v = [H_l(1) - H_l{}^h(1)] + [H_l{}^h(1+2) - H_l(1+2)]. \quad (87)$$

A Van der Waals fluid may be treated as a collection of hard spheres perturbed by the intermolecular attractions between the particles [58, 59]. On this basis the Helmholtz free energy of a liquid mixture is [60]:

$$A_l(1+2) = A_l{}^h(1+2) + RT \sum n_i \ln X_i$$
$$+ \frac{\rho n}{2} \int_{r_{ij}}^{\infty} u_{ij}(r) g_{ij}(r) 4\pi r^2 \, dr \sum X_i X_j \quad (88)$$

with ρ the solution density, $g_{ij}(r)$ the hard sphere liquid distribution function, $u_{ij}(r)$ the intermolecular potential, and r_{ij} the hard sphere excluded radius. Second and higher order perturbation terms that should be included in equation 88 have been assumed to be negligible. It has already been established that the entropy of apolar gas solutions is approximated by the entropy of the corresponding hard sphere mixed fluid; that is, the entropy term which corresponds to the integral on the right hand side of equation 88 is close to zero. Moreover, it should be recognized that $RT \sum n_i \ln X_i$ arises from the ideal mixing entropy, and does not contribute to the enthalpy. Therefore, the enthalpy of the liquid mixture is given by:

$$H_i(1+2) = H_i^k(1+2) + \frac{\rho n}{2} \int_{r_{ij}}^{\infty} u_{ij}(r) g_{ij}(r) 4\pi r^2 \, dr \sum X_i X_j. \qquad (89)$$

The term in equation 86 for recharging the pure solvent, $[U_l(1) - U_l^h(1)]$, is the negative heat of vaporization and is given by equation 76. Therefore substituting equations 76 and 89 into 87 yields:

$$\Delta H_2^v = -RT \left\{ \ln \left[\frac{V_g}{V_l(1)(1-\xi)} \right] + 3\left(\frac{\xi}{1-\xi} \right) + \frac{3}{2}\left(\frac{\xi}{1-\xi} \right)^2 \right\}$$

$$- 2\pi \rho n \int_{r_{ij}}^{\infty} u_{ij}(r) g_{ij}(r) r^2 \, dr \sum X_i X_j. \qquad (90)$$

And finally replacing $g_{ij}(r)$ with the hard sphere distribution function of a binary fluid mixture (equation 64):

$$\Delta H_2^v = -RT \left\{ \ln \left[\frac{V_g}{V(1)(1-\xi)} \right] + 3\left(\frac{\xi}{1-\xi} \right) + \frac{3}{2}\left(\frac{\xi}{1-\xi} \right)^2 \right\}$$

$$- 2\pi \rho n \sum X_i X_j \int_{r_{ij}}^{\infty} \frac{u_{ij}(r)}{1-\theta_3}$$

$$\times \left[r^2 + \frac{3}{2}\left(\frac{\theta_2 \sigma_i \sigma_j}{1-\theta_3} \right) r + \frac{3}{4}\left(\frac{\theta_2 \sigma_i \sigma_j}{1-\theta_3} \right)^2 \right] dr. \qquad (91)$$

The solute vaporization enthalpy may be evaluated with equation 91 provided the proper intermolecular potential function $u_{ij}(r)$ can be assigned. The explicit form of this function is, however, not known and is commonly approximated by a variety of expressions. Among these are the square well potential:

$$\begin{aligned}
U_{ij}(r) &= \infty & r &< r_{ij} \\
U_{ij}(r) &= \varepsilon_{ij} & r_{ij} &< r < a + r_{ij} \\
U_{ij}(r) &= 0 & a + r_{ij} &< r
\end{aligned} \qquad (92)$$

TABLE 11 SOLUTE VAPORIZATION ENTHALPIES[a]

Gas	CCl_4 (kcals/mole)		H_2O (kcals/mole)	
	$\Delta H_2{}^v$(calc)	$\Delta H_2{}^v$(obs)	$\Delta H_2{}^v$(calc)	$\Delta H_2{}^v$(obs)
He	−1.34	—	0.50	0.84
Ne	−0.77	—	1.02	1.89
Ar	−0.03	0.07	2.45	2.73
Kr	0.42	—	3.13	3.55
H_2	−0.97	−1.34	1.01	1.28
N_2	−0.56	−0.63	2.35	2.14
CH_4	−0.81	0.71	3.10	3.18

[a] The results have been taken from references 16 and 55.

and the Lennard-Jones [6–12] potential:

$$u_{ij}(r) = C_{ij} \sum [r^{-6} - r^{-12}r_{ij}] \tag{93}$$

where the constants ε_{ij}, a, and C_{ij} are taken as adjustable parameters or are calculated by standard procedures [49] from gas virial coefficients. Although the calculated enthalpies must be considered approximate, satisfactory results have been obtained. Boublik and Benson [57] adequately reproduced heats of mixing for cyclopentane-carbon tetrachloride solutions using equation 91, assuming a square well potential. For the dilute solutions of apolar gases in liquids, equation 91 may be further simplified because the mole fraction of the solvent is negligibly different from one. Thus that portion of the integral in equation 91 which involves solvent-solvent interactions becomes equivalent to the enthalpy of pure solvent vaporization and cancels with the bracketed term; equation 91 becomes:

$$\Delta H_2{}^v = 2\pi \int_{r_{ij}}^{\infty} \frac{u_{ij}(r)}{1-\xi} \left[r^2 + \frac{3}{2}\left(\frac{\xi\sigma_2}{1-\xi}\right)r + \frac{3}{4}\left(\frac{\xi\sigma_2}{1-\xi}\right)^2 \right] dr \tag{94}$$

Using a variant of equation 94, and assuming the Lennard-Jones [6–12] potential Pierotti [16, 55] calculated solute vaporization enthalpies for apolar gases dissolved in various apolar solvents, and water. Some of his results are compared with experimental values in Table 11. The correspondence is good enough to suggest that the Van der Waals approximation may be taken as a starting point for the calculation of apolar gas solution enthalpies with all solvents [133].

3.3 Further Applications of Scaled Particle Theory

In addition to the binary systems considered so far, scaled particle theory has been applied to more complicated solutions. We now consider briefly this extension of theory to the phenomenon of salting out. Because excellent reviews on the salting out of small apolar molecules [76] and the significance of these systems in terms of protein solutions [69] are available, the detailed discussions found therein will not be repeated here. The solubility of an apolar compound in water may be either increased or decreased by the addition of a second solute, the cosolute. Since the cosolutes most frequently studied have been salts, this phenomenon has been termed salting in (increased solubility), and/or salting out (decreased solubility). However, nonionic cosolutes, such as urea, also effect solute solubility [69] excluding a simple electrostatic explanation for the phenomenon. It has been observed empirically that at low concentrations of the third component the solubility of apolar molecules in water may be described approximately with:

$$\ln c_2^0/c_2 = K_s c_3 \tag{95}$$

where c_2^0, and c_2 are the saturated solute concentrations in pure water and in water containing a third component at the concentration c_3. K_s, the Setschenow or salting out constant, is dependent on both the solute and the cosolute. Recasting equation 95 in terms of vaporization free energies

$$\frac{\Delta G_2^{v,0} - \Delta G_2^v}{RT} = K_s c_3 \tag{96}$$

it is seen that positive values of K_s reflect salting out, a greater tendency of the apolar molecule to escape the aqueous solution and that negative values reflect salting in. At higher cosolute concentrations equation 95 frequently breaks down so that equation 96 will be taken as applying to dilute salt solutions only.

In order to determine the dependence of K_s on the nature of the cosolute it is necessary to explain the magnitude of ΔG_2^v as defined in equation 96. Various theories have been proposed. The earliest were based on electrostatic arguments only. As discussed by Long and McDevit [76] these were inadequate, for they could not encompass salting in, nor predict the behavior of nonionic cosolutes. The structural promotion theory of Frank and coworkers was extended by Frank and Franks [27] to explain both salting in and salting out. Equation 30, derived by these two authors for a two-species model of water, contains a term that depends on the relative amounts of the two species. If one of the two is a better solvent for the apolar molecule, requiring that $y \neq z$ (eq. 30), then the magnitude of the vaporization free energy will depend on the fractional amounts of the two waters. It was thus postulated that cosolutes affect the vaporization free energy of various compounds by shifting the position of the water structural equilibrium in one or the other direction. Because it had been proposed that nonstructured water is a poorer solvent than icelike water [11, 24, 27], it was further suggested that salting out was caused by an equilibrium shift towards the nonstructured water species and vice versa. In this way the qualitative concept of structure making, and structure breaking, by the cosolute is introduced. The proposal does not, however, have a quantitative basis. This has led to conceptual difficulties [77], not the least of which is the lack of criteria by which to define a compound as a structure maker or breaker.

McDevit and Long [78] have proposed that cosolutes could modify the solution properties of water by altering the internal pressure of this solvent. The internal pressure $(\partial U/\partial v)_T$ is the change in a system's internal energy during an isothermal volume change and, as seen from equation 70, has the same units as pressure. The more positive the internal pressure, the greater is the energy input required to separate the molecules of the system from one another. This thermodynamic parameter is a measure of the cohesive, attractive forces between molecules. There should be no confusion between the internal pressure and the isobaric expansivity (eq. 67). For example, the density of liquid water passes through a maximum as the temperature is increased through 4°C, requiring the isobaric expansivity to pass simultaneously through zero. The internal pressure of water near 4°C is, however, a large positive quantity, as reflected by the very large vaporization enthalpy. Conversely, $(\partial U/\partial V)_T$ is zero for a theoretical fluid of hard spheres by definition, yet such a fluid has a finite expansivity (eq. 68). If the liquid with the cosolute has a greater internal pressure than pure water, then more work would be required to form the cavity that contains the apolar solute. The cosolute would thereby salt out or decrease the solubility of the apolar solute. A decrease in internal pressure due to addition of a third component would have the opposite effect. On the basis of this concept an analytic expression for the Setschenow constant at limiting cosolute concentrations was derived [78]:

$$\lim_{n_2, n_3 \to 0} K_s = \frac{\bar{V}_2(V_3 - \bar{V}_3)}{\beta_1 RT} \tag{97}$$

with $\bar{V}_2$ and $\bar{V}_3$ the partial molar volumes of solute and cosolute in water, V_3 the hypothetical volume of the third component as a liquid, and β_1 the isothermal compressibility of pure water. Although the internal pressure theory of McDevit and Long was a marked improvement over earlier proposals, in that they were able to obtain an analytic expression that predicted both salting in and salting out, equation 97 results in values of K_s that are in poor quantitative agreement with experimental results, especially for larger solutes.

Recently Shoor and Gubbins [79] have applied scaled particle theory to calculations of apolar gas solubilities in salt solutions. Their calculations for argon in aqueous potassium hydroxide are presented in Figure 6 together

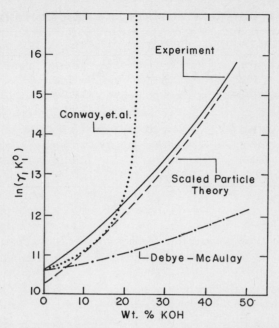

Figure 6   Theoretical and experimental activity coefficients of argon in potassium hydroxide solutions. This figure is reproduced from reference 79. The solid line represents the experimental curve. The Debeye-McCauley, and Conway et al. theories are electrostatic arguments.

with experimental results and values calculated with two electrostatic theories. The use of the scaled particle approach was expanded by Masterton and Lee [80], who computed Setschenow constants for a number of aqueous solutions. Their results for benzene are compared in Table 12 with experimental values and with results calculated on the basis of other theories. Scaled particle theory is seen to account for salting out better than either the McDevit-Long or electrostatic theories. Although Setschenow constants for systems displaying salting in have not yet been computed, there is qualitative evidence for the applicability of the scaled particle theory to these as well.

On the basis of equation 96:

$$\frac{\Delta H_2^{v,0} - \Delta H_2^{v}}{RT} - \frac{\Delta S_2^{v,0} - \Delta S_2^{v}}{R} = K_s C_3. \tag{98}$$

According to the Van der Waals approximation K_s will contain contributions from the hard sphere and attractive interactions between solvent, solute, and cosolute. If the hard sphere interactions are accounted for entirely by the

TABLE 12 SETSCHENOW COEFFICIENTS FOR BENZENE AT 25°[a]

Salt	Scaled Particle	Internal Pressure	Electrostatic	Observed
NaCl	0.188	0.42	0.166	0.198
KCl	0.162	0.34	0.156	0.166
NaBr	0.153	0.35	0.163	0.155
LiCl	0.124	0.31	0.172	0.141
RbCl	0.089	0.31	0.153	0.140
KBr	0.128	0.34	0.153	0.119
NaI	0.137	0.27	0.158	0.095
CsCl	0.085	0.26	0.150	0.088
CsI	0.034	0.11	0.141	−0.006

[a] Taken from reference 80.

entropy of the solution process, then the Setschenow constant can be broken down:

$$K_s = K_s^h + K_s^s \qquad (99)$$

with:

$$-\frac{\Delta S_2^{v,0} - \Delta S_2^{v}}{R} = K_s^h c_3. \qquad (100)$$

Lucas and de Trobriand [117] have calculated K_s^h for apolar gases dissolved in tetraalkylammonium salt solutions. As expected, the values they obtained were not identical with experimental Setschenow constants. Importantly, however, the variation of the computed K_s^h with salt molality parallels the variation of the experimental K_s (Fig. 7). Since $\Delta H_2^{v,0}$ must be a constant, this parallelism may be interpreted as arising from the linear dependence of the attractive interaction between solute and cosolute at low concentrations of the latter. This interpretation is entirely satisfactory, and we conclude that salting in by tetraalkylammonium compounds can be handled with the scaled particle approach. As an aside, the results presented here suggest that the salting phenomenon is a net result of two opposing tendencies. The introduction of the cosolute into the solution enhances the solubility of an apolar gas due to the Van der Waals attractions between the two. Simultaneously, the gas solubility will be decreased due to the greater solvent number density, $\rho_1 + \rho_3$, with the solvent considered as the water and cosolute taken together. Whether salting in or out is observed depends, therefore, on which of these tendencies predominates. This is, in turn, a function not only of the cosolute, but of the solute as well.

Scaled particle theory should be applicable to kinetic as well as to thermodynamic phenomena. A start in this direction has been made by McLaughlin [63], who calculated relative diffusion coefficients in binary mixtures.

The Thorne equation for mutual diffusion in a binary fluid of hard spheres is:

$$D = \frac{3}{2(\sigma_i + \sigma_j)}\left[\frac{kT}{2\pi}\left(\frac{m_i + m_j}{m_i m_j}\right)\right]^{1/2} \frac{1}{\rho g_{ij}(r)} \tag{101}$$

where D, m_i, and m_j are the mutual diffusion coefficient, and the masses of the ith and jth particle, respectively. The other symbols have been assigned previously. When the self-diffusion coefficient of the pure solvent is defined as:

$$D_1{}^0 = D(\sigma_i = \sigma_j; \ m_i = m_j) \tag{102a}$$

and the diffusion coefficient of the infinitely dilute solute as:

$$D_2{}^\infty = \lim_{n_2 \to 0} D \tag{102b}$$

then equation 101 can be rewritten:

$$\frac{D_2{}^\infty}{D_1{}^0} = \left(\frac{m_1 + m_2}{m_2/2}\right)^{1/2}\left(\frac{\sigma_1}{\sigma_1 + \sigma_2}\right)^2\left[\frac{2 + \xi}{(1 - \xi) + 3\sigma_2/(\sigma_1 + \sigma_2)}\right] \tag{103}$$

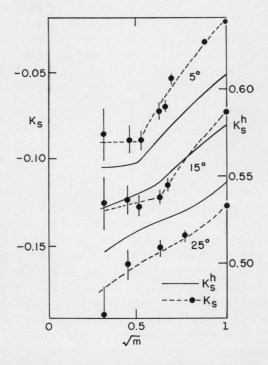

Figure 7 Propane salting in by tetrapropylammonium bromide. The solid lines are the hard sphere Setschenow constants calculated at various salt concentrations. This figure was adapted from reference 117.

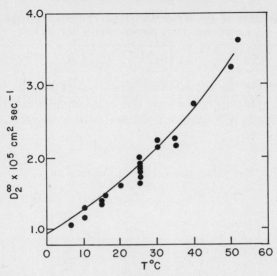

Figure 8   Temperature dependence of CO_2 diffusion in water; comparison of theory and observation. This figure was adapted from results presented in reference 63.

with the distribution function $g_{12}(r)$ obtained from equation 64. In equation 103 the solute diffusion coefficient has been expressed in terms of the solvent self diffusion. If $D_1{}^0$ is replaced with the observed self-diffusion coefficient of water, then the derived equation adequately reproduces the temperature dependence of CO_2 diffusion in water (Fig. 8). One may also compute the diffusion coefficients of solutes dissolved in carbon tetrachloride and water at one temperature. For most cases the calculated and experimental values agree within experimental error (Table 13). Tham and Gubbins have recently shown that the same procedure may be utilized to calculate relative diffusion coefficients in three-component systems [132].

4   APOLAR BONDING

4.1   Mechanism of Gas Dissolution

Utilizing the theoretical results of the preceding section, we now return to the consideration of two central problems that emerged earlier but could not be satisfactorily resolved; the validity of the structural stabilization theory of Frank and co-workers and the appropriateness of either the liquid-liquid, or the liquid-clathrate model as a description of the apolar bond.

The successful application of scaled particle theory to the liquid solutions of apolar gases suggests that these fluid mixtures may be treated as hard sphere systems perturbed by intermolecular attractions. The validity of this,

TABLE 13 DIFFUSION OF APOLAR SOLUTES THROUGH WATER AND CARBON TETRACHLORIDE AT 25°C

Solute	$D \times 10^5$ (cm²/sec) in H_2O		$D \times 10^5$ (cm²/sec) in CCl_4	
	Calculated[a]	Observed[b]	Calculated[c]	Observed[d]
He	5.03	6.3	—	—
H_2	6.27	4.8, 7.1	17.9	9.75
Ne	2.70	2.8(22.2°)	—	—
Ar	1.90	2.0(21.7°)	3.82	3.63
N_2	1.87	2.25	4.12	3.42
O_2	1.88	2.40	—	—
NO	1.78	2.57	—	—
CF_4	—	—	1.85	2.04
I_2	—	—	1.42	1.45
CH_3Cl	1.26	1.33, 1.45(22.1°)	—	—

[a] Calculated with the diffusion coefficient of H_2O^{18} in water 2.66×10^{-5} cm²/sec [64].
[b] All values were collected by McLaughlin [63]. For determinations made at a temperature other than 25°C, the experimental temperature is given in parentheses.
[c] Calculated with the self-diffusion coefficient of CCl_4, 1.41×10^{-5} cm²/sec [65].
[d] Values obtained from references 66 and 67.

the Van der Waals approximation, is not unexpected when the solvents are apolar, since scaled particle theory has also been applied successfully to pure apolar liquids. Because of the relatively strong and angularly dependent hydrogen bond interactions between water molecules, it is initially surprising that aqueous solutions can be mimicked in certain aspects by a hard sphere model. The minimum conditions one might invoke for the application of the Van der Waals approximation are weak, centro-symmetric attractive potentials. For pure liquids, in fact, the hard sphere approach does begin to break down in the case of hydroxylic solvents such as water and alcohols [46]. However, an argument to explain the success of the Van der Waals approximation for aqueous solutions can be readily formulated. The solute concentration in a saturated apolar gas solution is sufficiently low at atmospheric pressures that the partial molar volume of the solvent in the mixture does not differ significantly from the molar volume of the pure solvent, that is, the number density of solvent in the mixture is almost identical with that of the pure liquid. If, as would be true for apolar gases, the solute interacts only weakly with the solvent, then small amounts of the second component could be fitted into available liquid cavities with little structural disorientation; in other words, the dissolution of apolar gases would constitute a packing problem. This reasoning is supported by the differences in partial molar solute volumes observed for apolar gases dissolved in organic liquids versus water. The smaller values in aqueous solutions can be interpreted as arising from the smaller molar volume of liquid water coupled with its more open structure. If the solution of a dilute apolar solute were properly described as a packing phenomenon, with no gross disorientation of the solvent structure, then the solvent-solvent interaction of the pure liquid would be retained in the mixture and the solvent-solute interaction would become the predominant factor governing the solution thermodynamics. Because the interaction of an apolar solute with the surrounding solvent is weak and centro-symmetric, the solution process would be approximated by the perturbed, hard sphere model.

This argument and its conclusion do not fit within the framework of the theory proposed by Frank et al. that the solution of apolar gases in water induces gross structural alterations in the solvent. There is, on the other hand, a qualitative overlap between the viewpoint offered here, and the theory of Eley, who treated aqueous apolar gas solutions as a problem in molecular packing. We have as support for this argument the entropy and enthalpy calculations of the last section, which are successful for water, carbon tetrachloride, and other liquids [55, 133] and which do not require the introduction of a special mechanism for aqueous solutions. The formation of aqueous and nonaqueous solutions appears to be basically the same, and should solute stabilization of icelike water occur, it would contribute little to the observed thermodynamic parameters. Stated differently, we conclude that the magnitudes of the thermodynamic parameters for gas solution in water do not constitute evidence for a theory of structural stabilization, even though the theory was first proposed to explain these values [11]. The possibility should not, however, be discarded that the packing process defined by the equations of the preceding section is, in fact, similar to the structural stabilization theory of Frank and co-workers, even though the formulations differ. Were this the case, then the scaled particle approach would be the preferred formulation, since this theory yields simple, analytic expressions for the thermodynamic parameters of solution formation. The empirical nature of the calculations, which are based on the scaled particle equation of state, should be emphasized; the experimental densities of the pure solvents must be inserted into the derived equations. This cautionary note, however, further emphasizes that the apparent uniqueness of water as a solvent is derived from the structure of that liquid, and not from a unique dissolution process; for it is the molecular structure which determines the density of the pure liquid.

That the equations derived for binary solutions can be extended to describe salting in and salting out in three-component systems and that the scaled particle approach can be utilized to calculate relative diffusion coefficients further support the position that a structural stabilization mechanism is not required in order to understand aqueous solution properties.

Thus the smaller diffusion coefficients of apolar compounds in water as compared with organic liquids could be ascribed to the presence of an icelike structure surrounding the solute, thereby increasing its effective bulk. However, the applicability of equation 103 to aqueous solutions implies that apolar solutes do not, in fact, diffuse abnormally slowly if the standard taken for comparison is the self-diffusion of the solvent molecules.

Finally, there are two qualitative arguments that suggest that no special mechanism need be invoked to explain the thermodynamics of gas dissolution into liquid water. The first of these has already been presented and is based on the observation that relatively large entropies and enthalpies are observed for solutions of apolar gases into preformed water lattices (Table 7). In these processes no lattice stabilization can be inferred, since maximum structuring occurs before the introduction of solute. Thus the large vaporization enthalpies must arise from Van der Waals interactions between lattice water and the solutes. This suggests that the similarly large enthalpies obtained with liquid water solutions arise from solute-solvent and not from solvent-solvent interactions. At the very least, we may conclude that the magnitudes of solute vaporization enthalpies and entropies cannot be evidence *per se* for a structural stabilization mechanism. The second argument concerns the free energy of the solution process. The very low solubility of apolar gases in water, when compared with organic liquids, has been an important factor behind the proposal for a special dissolution mechanism. If a unique mechanism were not operative in aqueous solutions, then it would have to follow that the low solubility of apolar compounds is not abnormal. Uhlig [61] proposed that the logarithm of the Ostwald coefficient (a measure of gas solubility) for a single apolar gas should vary linearly with the surface tension of the liquid solvent in which it is dissolved. Although the theoretical argument on which the proposal was based is no longer accepted, the linear relationship has been experimentally verified. It is reasonable to expect that a solvent with a unique dissolution mechanism would not obey the same Uhlig relationship obtained for other solvents. Specifically, if the low solubility of apolar gases in water were abnormal, then the Ostwald coefficients obtained with water should be lower than predicted on the basis of a relationship constructed with organic solvents. This is not true for argon, which yields a reasonably linear variation with 25 solvents ranging in polarity from hexane to water [62] (Fig. 9). We conclude that the low solubility of argon in water is not abnormal and does not arise from a unique solution process, but rather is a reflection of those same molecular forces that dictate the liquid's surface tension.

In summary, we believe that for *dilute* solutions of apolar molecules the properties of liquid water as solvent can be satisfactorily explained without the introduction of a special solution mechanism. It should be emphasized

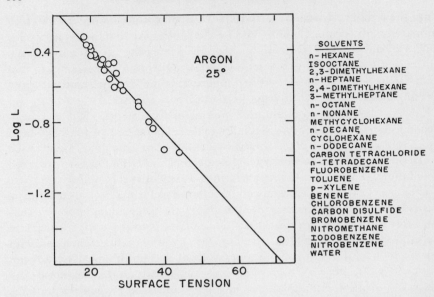

Figure 9   Dependence of Ostwald coefficients of argon on solvent surface tension. Reproduced from reference 62.

that the structural stabilization mechanism cannot be eliminated definitively with the arguments presented here; however, elimination of the additional assumptions required by this theory is preferable on the basis of logical economy. There is support for lattice stabilization when we consider concentrated solutions of more polar compounds that dissolve well in water. Glew et al. [68] have reviewed various properties of aqueous solutions containing hydrophilic organic compounds such as acetone. The variations of these properties with solute concentration often pass through characteristic minima or maxima (Table 14). The solution compositions at these extrema correspond to the compositions of the crystalline clathrates formed with these solutes at lower temperatures. In explanation, these authors propose that the organic molecules stabilize an incompleted clathrate structure containing approximately the same number of water molecules as found in the solid hydrate. Some of the earliest evidence cited in support of a structural promotion theory was based on the thermodynamic and transport properties of aqueous solutions containing organic ions such as the tetraalkylammonium cations [24]; this topic has been reviewed thoroughly [69] and is not treated here once again. Perhaps the most convincing evidence for the lattice stabilization concept is the formation of crystalline hydrates around organic guest molecules as described in an earlier section. If the available data obtained with these more concentrated solutions can be interpreted in

TABLE 14 WATER SOLUTE RATIOS FOR HYDRATES AND SOLUTION PROPERTY EXTREMA[a]

	Ethylene Oxide	Dioxane	Acetone	Tetrahydro-furan	t-Butyl Alcohol
Hydrate coordination number	20–24	28	28	28	28
H_2O excess molar volume	29	28	27	29	24–32
H_2O activity coeff. minimum	24	25	—	—	26
PMR chemical shift maximum	17	17	17	17	16
Apparent $\bar{V}_2$ minimum	19	18	15	17	20
Liquid compressibility minimum	—	14	10	—	30
Sound velocity maximum	—	10	16	—	10–24

[a] This table was taken from reference 68.

terms of the theory of Frank et al., then we must question whether such a solution mechanism holds for concentrated but not dilute solutions, whether it applies for both contradicting our own previous conclusions, or whether there is a common solution mechanism that may be formulated in terms other than those of the structural stabilization theory.

It is intellectually economical, as formalized by Occam's razor, to assume that the physical properties of dilute and concentrated aqueous solutions have common physical bases. With this assumption it could be argued that the success of the scaled particle approach in reproducing the properties of dilute solutions is fortuitous, and misleading, because apolar stabilization of icelike water is observed at higher solute concentrations. However, the scaled particle approach is so simple that this reviewer would be loath to dismiss it, if only for practical reasons. Moreover, the properties of dilute solutions that can be reproduced reasonably by the theory are sufficiently extensive so as to make a fortuitous correspondence appear unlikely. If the scaled particle approach together with the Van der Waals approximation is applicable to dilute solutions, then, as was previously concluded, the properties of these solutions can be envisioned as arising from molecular packing processes that occur in a volume fixed by the intermolecular solvent attractions. Could the stabilization of water clathrates, in particular the induced liquid to solid transition of crystalline hydrates also be explained as a packing phenomenon? Perusal of the equations of state derived by the scaled particle theory (eqs. 60 and 62) reveals that these cannot yield the discontinuities required for the description of a phase transition. This inadequacy is not due to the hard particle model itself, but rather to the inadequacies of the assumptions used in obtaining the analytic form of the contact radial distribution function $g(r)$. Computer simulations of hard sphere fluids do exhibit phase transitions [70], and anisotropic to isotropic transitions may be obtained from a scaled particle treatment of cylindrical [118] and rectangular [119] particles. Because of their inadequacy, equations 60 and 62 cannot be utilized for a quantitative answer to the question just posed.

A qualitative argument describing clathrate formation as a packing process is possible. The results of x-ray crystallography reveal the large number of different lattice structures water may assume [28, 29]. These different structures have different water densities depending on the sizes of the cavities within the lattice. For example, van Cleef and Diepen [71] have concluded from thermodynamic data that conversion of ice I to the type-I hydrate is accompanied by a volume increase of 15 ml/mole. The energy required to stabilize a lower density lattice cannot be obtained by the formation of new hydrogen bonds, because maximal bonding is observed in all water lattice structures [29, 30]. Clathrate stability appears to derive from occupancy of the newly formed cavities by solutes that do not compete successfully with water in hydrogen bond formation and that thus fill the void without destroying the surrounding lattice. For this reason the stabilization of a type-I or type-II clathrate by an apolar guest molecule depends primarily on the size and geometry of the guest [28]. At very low concentrations a weakly interacting solute could dissolve into liquid water by occupying available cavities or structural defects leaving the original solvent structure essentially intact. At some higher concentration, however, the previously available holes will have been filled. In order then to add more of the second component, the solvent number density must decrease, because the energy required to introduce additional solute without simultaneously introducing more cavities becomes greater than the energy required to alter the liquid structure. Because water can assume a variety of low-density, hydrogen-bond-stabilized lattice networks, the most energetically favorable volume increase could occur with a minimum number of hydrogen bond disruptions. If a lower-density, isohydrogen-bond structure accommodating more solute molecules were available, then this structure would be formed. It is, therefore, reasonable to argue that clathrate stabilization by a weakly interacting solute is due to packing restrictions imposed by the water. The same type of stabilization would not be possible with apolar solvents, which lack the relatively strong, angular hydrogen bond and cannot construct the appropriate clathrates.

Although the formulation of clathrate stabilization as a packing phenomenon is a qualitative and speculative argument, it should be recognized that the flickering cluster formulation is no less qualitative or speculative. If a physical process may be simultaneously explained with more than one reasonable argument, it is safest to conclude that the process is not well understood. Stabilization of water clathrates at higher concentrations of organic solutes apparently contradicts the application of scaled particle theory to dilute aqueous solutions only if clathrate stabilization is explained with the theory of Frank et al. If such stabilization is visualized as a result of packing restrictions imposed on the solute by water, there is no contradiction. This latter visualization is an extension of Eley's proposal explaining the thermodynamics of aqueous apolar gas solutions. As was previously noted, the theories of Eley and of Frank et al. are not mutually exclusive.

4.2 The Protein Interior

We turn from the brief discussion on the mechanism of the solution process for mixing apolar solutes and water to the closely related question of what constitutes an adequate description of the apolar bond. For even though apolar bonding is widely acknowledged to be important in the thermodynamic treatment of proteins, there is as yet no precise definition of this bond. Instead, it is described in terms of model systems, either the liquid-liquid or the liquid-clathrate partition arguments of Kauzmann and Klotz. Thus we must first determine whether either of these thermodynamic models adequately describes apolar interactions within proteins before a quantitative discussion on the importance of the apolar bond can be attempted. In both the Kauzmann and Klotz models freely moving, small molecules are partitioned between an organic liquid and an aqueous phase, which is either a liquid or a preformed clathrate lattice. Even though both models have been associated with the structural stabilization theory of Frank and co-workers, the molecular mechanism underlying the partitioning is not of primary concern, since we require from the model systems only the thermodynamic parameters that characterize the partition process. If the free energies and other quantities obtained with the models do apply to proteins, then the thermodynamics of apolar bonding are established. Thus, the adequacy of any particular partitioning model as a description of apolar interactions within proteins rests solely on the validity of the assumptions required to apply that model to the macromolecular system. These assumptions are that apolar side chains are similar to small, apolar solutes; that a protein interior is properly described as a liquid; and that water at the protein surface is similar to either liquid or clathrate water.

All these assumptions cannot be completely valid, which is self evident. What must be determined, therefore, is how closely these premises approach reality. This is a quantitative problem that requires the protection of a quantitative theory for a satisfactory attack. But it is the lack of such a theory that necessitates an appeal to model systems, and we appear to have entered a maze with an entrance but no exit. An additional difficulty arises from the nature of the experimental data upon which any theoretical discussion must be based. There are three sources of thermodynamic results to which we may turn: protein denaturation, polypeptide chain aggregation, and ligand binding. Excellent data are now becoming available from studies of protein denaturation [4, 13]. Chain unfolding, however, involves alterations of more than a single bonding type. The attempt to separate the various thermodynamic contributions to the net, observed results remains an empirical exercise, so that the thermodynamic parameters experimentally assigned to apolar interactions have a problematic significance for further discussions. Experimental efforts to determine the thermodynamics of protein subunit aggregation have been stimulated recently by theoretical, computational, and technical advances. Nonetheless, the information generated by this activity has been meager, with the enthalpy and entropy data required for meaningful discussion lacking in most cases [92]. In contrast, studies on the binding of small ligands to proteins have engendered a large store of results. However, in most cases the apolar interaction contributes in part to the overall process, polar interactions being involved simultaneously. Thus, as in the case of denaturation results, parameter separation again becomes a problem.

An apparently reasonable approach to the isolation of the apolar interactions between ligand and protein has been to vary that part of the small molecule that is assumed to be responsible for the interaction and then to compare protein-ligand affinity variations with the partitioning of the same ligand series between water and an arbitrary organic solvent. In two examples of this procedure [108, 109, 110] the observed ligand affinities and phase partition constants varied linearly with respect to each other. This linearity was then interpreted in terms of apolar bonding. Two reservations concerning this line of reasoning should be considered. First, free energy changes are relatively insensitive indications of a solution mechanism due to entropy and enthalpy compensations (Tables 1 and 3). Second, ligand binding to macromolecules can be considered as a partitioning process, a point discussed by Canady and co-workers [134]. Comparing the tendency of a solute to escape liquid water into either a protein binding site or an organic solvent is equivalent to discussing two partitioning processes involving one common phase. This is merely a comparison of the solute solubility in the protein versus the organic liquid, and the observed linear relationships mean that the free energy of the solute dissolution into the organic liquid is linearly proportional to the free energy of the solute 'dissolution' into the protein; that is, the water contribution has been eliminated. The significance of such a proportionality for clarification of the concept of apolar bonding is unclear. It should also be recalled that when the solubilities of organic compounds are compared in various solvents, the solubility of an apolar molecule in water is roughly independent of the molecular size at 25°C (Table 3). (Canady has pointed out in a personal communication that a similar independence is observed in the series of derivatives benzene, toluene, xylene, ethylbenzene, and cumene.)

An approach of interest is the measurement of apolar gas adsorption to proteins. This process should yield thermodynamic parameters free of the complications that arise from ionic or permanent dipole interactions between the ligand and the protein. Moreover, the solution of gases into a variety of solvents has been studied sufficiently to provide a large amount of the data required for empirical and theoretical comparisons. Proteins differ from liquids as "solvents." For liquids Henry's law is approximately obeyed over the range of gas pressures commonly examined. In contrast, proteins interact according to Langmuir isotherms [93, 94] indicating gas adsorption at discrete sites on the protein. In this respect proteins resemble the crystalline clathrates that hold apolar solutes within discrete cavities and also display a pressure independent saturation [28, 34]. There are additional resemblances between these two apparently different types of solvent. Thermodynamic quantities describing gas-protein interactions are presented in Table 15. The results taken from the literature have been recalculated as vaporization energies in order to focus on the solute interaction with the protein by eliminating the contributions of the aqueous phase. The magnitudes of the Gibbs free energies for xenon and butane vaporization suggest that proteins are generally better solvents than water and are comparable to or better than carbon tetrachloride. (As pointed out by Wishnia [94] the solubilities of gases in proteins appear to be even anomalously high in some cases.) By this criterion proteins would be considered similar to organic liquids as originally proposed by Kauzmann. In the case of xenon, however, the observed enthalpies and entropies are much greater than would be predicted for carbon tetrachloride. The limited results available preclude a detailed discussion, and the simple comparison of thermodynamic data is of dubious value because of possible energy contributions derived from changes of protein structure accompanying the adsorption process. For this reason the reaction of xenon with myoglobin is of particular interest. The results of x-ray crystallography [96, 97] indicate that xenon is bound to the heme protein within a cavity next to the prosthetic group. (Although more than one xenon is bound, this site is believed to hold the most firmly attached gas atom, for which the data of Table 15 apply.) Xenon appears to fit into the cavity with little or no alteration of the protein structure. In support of this conclusion Lee and Richards [115] have calculated from the atomic co-ordinates of myoglobin that the protein has an empty cavity at this site large enough to contain a xenon atom. Thus the binding of xenon to myoglobin is analogous to the adsorption of gases into crystalline clathrates, in that both involve the occupation of a preexistent cavity with little or no structural distortion. For further comparison the entropies and enthalpies

TABLE 15 VAPORIZATION FROM LIQUID WATER, CARBON TETRA-CHLORIDE, HYDROQUINONE CLATHRATE AND PROTEINS

Gas	Solvent[a]	ΔH_2^v (kcal/mole)	ΔS_2^{vb} (e.u.)	$\Delta G_2^v (\Delta A_2^v)^g$ (kcal/mole)
Xenon	H_2O	4.4	36	−6.3
	CCl_4	$(2.9)^c$	$(16)^c$	$(−1.9)$
	Mb^d	9.5	36	−1.2
	Mb^{+d}	11.6	30	2.7
	$MbCN^d$	13.4	24	6.5
Butane	H_2O	6.3	43	−6.5
	CCl_4	$(7.2)^c$	$(22)^c$	(0.6)
	Mb^{+e}	6.3	25	−1.2
	Hb^{+e}	7.3	26	−0.5
	β-LGA-1^e	8.7	29	0.1
	β-LGA-2^e	9.2	31	0
Argon	HQ	6.0^f	25^g	−0.7^g
Oxygen	HQ	5.5^f		
Nitrogen	HQ	5.8^f	26^g	−1.0^g
Hydrogen chloride	HQ	9.2^f	25^g	2.7^g
Hydrogen bromide	HQ	10.2^f		
Methanol	HQ	11^f		
Formic acid	HQ	12.2^f		

[a] Abbreviations: Mb–deoxymyoglobin, Mb+—metmyoglobin, MbCN—cyanometmyoglobin, Hb+—methemoglobin, β-LGA-1—β-lactoglobulin A monomer at pH 2.0, β-LGA-2—β-lactoglobulin A dimer at pH 5.3, HQ—hydroquinone β clathrate.

[b] Entropies and free energies are given on the unitary scale.

[c] Entropies were calculated from equation 83. Enthalpies were calculated from the empirical Barclay-Butler relationship presented by Frank for organic solvents [95]: $\Delta S_2^v = 12.75 + 0.00124\Delta H$.

[d] Data obtained from Ewing and Maestras [93]. The literature results were recalculated to yield the changes obtained in going from gas adsorbed on the protein to gas in the vapor state; e.g.: $\Delta H_2^v = \Delta H_2(\text{prot.} \rightarrow H_2O) + \Delta H_2^v(H_2O \rightarrow \text{gas})$.

[e] Data obtained from Wishnia [94]. The results were recalculated as described in the previous note.

[f] Data obtained from Evans and Richards [120].

[g] The values presented are the Helmholtz free energies reported by Van der Waals and Platteeuw [34]. Entropies were calculated with the internal energies provided by these authors.

of gas vaporization from the crystalline clathrate of hydroquinone are also presented in Table 15; the magnitudes of the clathrate and protein parameters are comparable, further suggesting an analogy between the two systems. Any similarity between gas adsorption to proteins and crystalline clathrates must, however, be considered cautiously. The supportive data are scanty. In addition, the binding of xenon to myoglobin depends on the nature of the ligand at the sixth coordination position of the iron, which is on the opposite side of the heme group. This may indicate the importance of specific interactions which cannot be interpreted in the simple terms of cavity occupation.

Faced with no quantitative theory and with insufficient experimental results, is it still possible to determine whether the partition models of Klotz and Kauzmann are valid in the description of protein apolar interactions? Or more specifically, can we determine how closely the assumptions on which these models are based correspond with physical reality? The ensuing discussion is offered in the spirit of skepticism to suggest that a partial answer may be feasible. One element of a polypeptide chain imposes both covalent and noncovalent restrictions on its neighbors. Because a small molecule dissolved in a liquid is not similarly restricted, amino acid side chains cannot be identical with a small solute. The degree, and hence importance, of this difference is unknown. Nor can much be said about the water in contact with the protein surface. Insofar as macromolecular hydration can be experimentally detected [e.g., 82–87], the water near the protein cannot be identical with the bulk liquid. This leaves open the possibility for clathrate structuring. However, the charged residues on the protein surface would tend to destabilize possible clathrates by incorrectly orienting the water molecules, and there is little support for ordered clathrates near the protein in the published results of x-ray crystallography.

The last assumption, the protein interior resembles an organic liquid or "oil drop," may be examined in more detail. Scaled particle theory cannot be presently applied to the consideration of side chains anchored on the polypeptide backbone, but a qualitative extension of this approach is feasible. On the basis of the equations derived in the previous sections, we see that various thermodynamic parameters characterizing the solution process in liquids are functions of the term ξ, the fraction of the total volume actually covered by the solvent molecules (equation 51′). If these equations are appropriate and if solute dissolution does devolve around the problem of packing, then introduction of a foreign molecule will be dependent on the amount of available space, a quantity reflected by the magnitude of ξ. (Although the theory of solute dissolution into crystalline clathrates [34] has not been described in this review, a similar conclusion may be drawn in that case as well.) By this line of reasoning we are led to the suggestion that, as a minimum requirement, the protein interior may be considered to resemble an organic liquid if the fraction of space covered by the atoms in the two are similar. For liquids this fraction may be estimated with equation 51′. For proteins it can be calculated provided the total protein volume and the volumes of the individual atoms are known [88]. Assuming atoms are hard, atomic volumes may be assigned from estimates made with small molecules [89]. Known amino acid sequences yield the total number of each atom type; thus the total volume covered by all the atoms may be computed. The volume of the folded protein can be determined in principle from x-ray crystallographic data; however, sufficient information is not generally supplied in the literature. Protein volumes may also be estimated from partial specific volumes; this procedure introduces possible errors, the consequences of which have been considered [88]. The parameter ξ calculated for six proteins from 14,000–150,000 daltons are presented in Table 16.

TABLE 16   FRACTION OF OCCUPIED SPACE IN THREE LIQUIDS AND SIX PROTEINS[a]

Compound[b]	ξ
Lysozyme (14,190)	0.766
Trypsinogen (23,998)	0.737
Chymotrypsinogen B (25,751)	0.745
Carboxypeptidase (36,381)	0.718
Glyceraldehyde-3-P dehydrogenase (145,523)	0.752
γ-Immunoglobulin Gl (150,740)	0.751
Average	0.747 ± 0.007
Water	0.363
Cyclohexane	0.439
Carbon tetrachloride	0.438
Closest packed spheres[c]	0.705

[a] Values obtained from reference 88.
[b] Molecular weights calculated from the primary structure are given in parentheses.
[c] Calculated on geometric considerations for hexagonally packed spheres of identical size.

These values are reasonably constant and remarkably high, averaging near 0.75. In comparison, ξ for cyclohexane and carbon tetrachloride is 0.44, and for a collection of closest hexagonally packed spheres of identical size 0.705. The magnitude of ξ obtained with six proteins suggests that these macromolecules might be better described as glasses rather than as mobile liquids, that is, as wax balls instead of oil drops. The concept that the protein interior is rigidly structured because of close atomic packing is compatible with the results of x-ray crystallography. That amino acid residues may be specifically located by a diffraction technique implies that the thermal movement characteristic of fluids is extensively reduced or eliminated within the protein. Space-filling protein models constructed with the atomic coordinates obtained from x-ray crystallography in fact display closely packed structures [41, 90, 125]. Stryer [41] has pointed out that several proteins have identical structures in different crystal lattices where intermolecular forces presumably differ. This suggests that the protein structure is not readily deformed, due perhaps to rigidity. Rupley's conclusion [107] that proteins, for which the comparison has been made, have similar structures in solution and crystal may also be interpreted in terms of structural rigidity.

The values of ξ calculated for proteins (Table 16) will be in error to the extent that the partial specific volume, a thermodynamic parameter, differs from the true protein volume. A better procedure for estimating the fraction of space covered within the protein would be to consider electron density distributions determined from crystallographic data. Although this approach has not yet been taken, Lee and Richards [115] have performed related calculations from which the number and approximate sizes of empty cavities inside proteins have been obtained. Their results (Table 17) are presented in terms of the volume available to the center of a 2.8 Å diameter sphere, an approximation for water. Ribonuclease S and lysozyme contain

TABLE 17   PROTEIN CAVITIES[a]

Protein	Cavity	Effective Volume (Å^3)
Ribonuclease S	A	0.029
	B	0.375
	C	0.0003
Lysozyme	A	0.017
	B	1.327
	C	0.01
	D	0.017
	E	0.002
Myoglobin	A	0.853
	B	0.0002
	C	0.007
	D	1.781
	E	0.470
	F	0.0002
	G1	0.371
	G2	0.178
	H	1.754
	I	2.229
	J	2.854
	K	0.016
	L	0.031

[a] Data taken from reference 115. The effective volumes are calculated as the volume accessible to a 2.8 Å diameter sphere. Thus a cavity of volume of 0 Å^3 can contain such a sphere, which is not free to move. Cavity A of myoglobin is the xenon binding site [97].

three and five such cavities, respectively. These are all small; because a 2.8 Å diameter means a volume of approximately 11 Å^3, the largest cavity in Table 17 is smaller than 15 Å^3. Comparing the sum total of the cavity volumes with the overall protein sizes, approximately 16,000 Å^3, we may again conclude that the polypeptide chains must be compactly arranged. The greater number of cavities within myoglobin is of interest; the high helical content of this protein may preclude the packing efficiency attained in proteins with more random coil. The cavities "seen" by Lee and Richards are sufficiently large to hold at least one water molecule empty, but are empty with the possible exception of three sites in lysozyme [121] and the water attached to the iron of metmyoglobin. Is it reasonable that holes left behind after packing the polypeptide chain should remain empty? The transfer of a water molecule from the bulk liquid into a protein cavity can be considered as a two-step process, vaporization of water from the liquid followed by solution of the gaseous water into the protein. The vaporization free energy of water at 25°C is approximately 2 kcal/mole [122]. There is no experimental estimate for the solution of gaseous water into an organic cavity. It is, however, reasonable to expect that this process would be less spontaneous than the reverse, solution of an apolar gas into a cavity built of water molecules. On this basis and using the data of Tables 6 and 7, the estimated free energy of water solution into a protein cavity is 1 kcal/mole or greater at 25°C. Thus a very crude value for the overall transfer of one molecule from bulk water to protein interior is 3 kcal/mole or greater. Although not large, this unfavorable free energy would result in less than 1 % occupancy of the available cavities, and these would appear empty by x-ray analysis.

Compact protein structures are commonly acknowledged by biochemists. However, schizophrenic molecules have been constructed by simultaneously visualizing a liquid interior. Consequently, the thermodynamic and mechanistic implications of tightly packed and rigid macromolecular structures have not been considered seriously. (For one exception, see reference 112.) An important implication for this discussion is that the liquid-liquid partition model is an inadequate representation of apolar interactions within proteins; that is, if the protein interior is a wax ball, and not an oil drop, then the thermodynamic parameters that describe apolar molecules partitioning between water and organic liquids are not applicable to a discussion of protein folding. The possible inadequacy of the liquid-liquid partition model has been discussed by Brandts and co-workers [91]. They have pointed out that volume changes that accompany protein denaturation are close to zero. Since denaturation is the net result of many, incompletely understood physical processes, experimental isolation of the volume contribution made by each is not feasible. These authors, nonetheless, estimated the magnitudes and signs of all possible contributions and then compared calculated and experimental results. On the basis of the liquid-liquid partition model the transfer of an apolar side chain from the protein interior to the aqueous environment would yield a negative volume change (Table 2, Fig. 5). Other contributions (e.g., the increased exposure of charged groups and peptide bonds) should also be negative. As a result, the total volume change predicted would be large and negative. From the very small denaturation volumes observed, Brandts and co-workers concluded that either a hitherto undescribed process, characterized by a large positive volume change, must be invoked or that the Kauzmann model is incorrect. If, however, the protein interior is closely packed, then exposure of apolar residues to the aqueous environment will contribute a positive volume change [88]. Were all other contributions negative, the observed changes could be explained without invoking an unknown process.

Observations concerning the binding of small ligands to proteins may also be interpreted in a new light, if the wax ball model of the protein interior is adopted. A rigid, or highly viscous structure is compatible with the empty cavities found within proteins, whereas a mobile structure is not. It would be reasonable to postulate that no large structural alterations would be observed when a small molecule is bound within a protein cavity or deep cleft. X-ray results support this conjecture. The binding of xenon to myoglobin [96, 97] has been mentioned already. Greer [129] has reported that 20–30% of the β-chain heme groups are missing in crystals of deoxyhemoglobin Hyde Park M. Although the side chains of those residues adjacent to the empty packets appear to relax slightly into the available space, no major structural changes are observed in the protein. Smiley and co-workers [129] have compared the apoenzyme of dogfish lactic dehydrogenase M_4 with the abortive ternary complex between enzyme, NAD, and pyruvate. Except for the position of a polypeptide loop on the proteins surface, which folds across the filled active site, the structures of the two derivatives are similar. The binding of small molecules to proteins may, therefore, be analogous to the binding of small molecules within crystalline clathrates. The thermodynamics of xenon binding to myoglobin has already been discussed in this respect. Glazer [111] has suggested that the interaction of aromatic dyes with enzymes occurs primarily at, or near substrate, prosthetic group or coenzyme sites; this has been observed by Wassarman and Lentz in the complex between tetraiodo-fluorescein and dogfish lactic dehydrogenase [136]. Since available x-ray results show all active sites as either clefts or pockets, the formation of an aromatic ligand-protein complex may have the thermodynamic requirement of an empty cavity. The work required to insert a molecule into the matrix of a tightly packed protein at a site where no hole exists could be too energetically expensive. Conversely, an empty cavity maintained within a protein need not be an energetic disaster due to the packing restraints imposed by steric interactions. It is well known that certain proteins, such as the serum albumins, bind organic molecules well, but others do not. The explanation for this cannot rest on the presence or absence of apolar regions within the protein, because we now expect all proteins to contain an apolar core. What may differentiate the good from the poor binding proteins is the availability of cavities large enough to contain the organic ligands.

Because of the magnitude of the estimated transfer free energy, it was previously held to be reasonable that the cavities described by Lee and Richards (Table 17) contain no water. It would for the same reason be reasonable for active sites built as deep, rigid clefts to contain no water either. The number of protein structures determined by x-ray crystallography is not sufficient to verify this conjecture. Most of those proteins that bind small molecules, and for which three dimensional structures are available, are hydrolytic enzymes; the exceptions are the oxygen binding heme proteins and lactic dehydrogenase. The active sites of the hydrolytic enzymes appear to be located on the protein surface or within a shallow groove exposed to the aqueous environment [90, 99, 123–127]. This may be because of the constraints imposed by the large substrates on which these enzymes operate. There may, however, be an additional thermodynamic requirement for exposure of the active site to the solvent, so that the water molecule acting as a substrate can reach the reaction center in sufficiently high concentrations. The structures of nonhydrolytic enzymes with water as one substrate will constitute an interesting test for this latter possibility. In contrast, hemoglobin, myoglobin, and lactic dehydrogenase have active sites within deep clefts or holes [103, 128]. These appear to be empty in the absence of the molecules those sites are meant to contain. The construction of an empty active site could be of functional importance since elimination of a required desolvation step might increase both the affinity of the substrate, and its rate of binding. In an aside we may note that as part of his argument for the theory of induced fit, Koshland [130] cited the low reactivity of water towards certain enzyme intermediates, which in turn react readily with the hydroxyl group of their substrates. Since he assumed that water would occupy the active site, the inefficient catalysis of a hydrolytic reaction in the absence of substrate was ascribed to an improper enzyme conformation. It may, however, be possible that the low observed reactivity is due in part to a low concentration of water in the active site cavity.

4.3   The Rigid Protein Interior—A Mechanistic Consequence

The concept of an enzyme maintaining, by its internal rigidity, an immobile, empty cavity for an active site is based on incomplete evidence and should be adopted hesitantly. Nonetheless, there are interesting consequences stemming from such a picture, which should be considered seriously. In the previous section thermodynamic implications are discussed briefly, here we turn to enzyme catalysis mechanisms. That a rigid structure can control substrate binding specificities through a lock-and-key interaction is evident. Koshland [130] has argued that the lock-and-key mechanism cannot also account for rate enhancement specificities, that is, for the ability of an enzyme to discriminate between closely related compounds by controlling the maximum velocity of substrate turnover. Rate enhancement specificity can, however, be explained by his proposed theory of induced fit. As originally presented, this theory raised the image of major conformational alterations during a catalytic cycle. In a tightly packed protein large conformational changes would require extensive chain unfolding and refolding. In graphic terms, the more closely the protein interior resembles a wax ball, as opposed to an oil drop, the greater will be the "internal viscosity"; hence, the larger the activation energy required to change the position of a residue relative to its neighbors. Thus a major change in secondary or tertiary structure as an obligatory step in an enzyme catalyzed reaction could be energetically expensive, tending to slow the overall rate. In fact, conformational rearrangements at the secondary or tertiary levels are relatively minor where seen in x-ray crystallographic results, with the exception of changes involving chain segments located at the protein surface [41, 99–102, 129]. These observations do not, however, contradict the basic premises of the induced fit theory, because apparently minor shifts in key catalytic groups may be envisaged as conferring the proper specificity [137].

That the concept of a rigidly structured protein is compatible with an induced fit mechanism is a negative conclusion. In a positive sense, the rigid active site may be used in support of other mechanisms. The possible importance of a rigid structure as a corollary of the strain mechanism of enzyme catalysis has been discussed by Jencks [3]. In one formulation of this theory, the energy gained from substrate binding to the protein may be utilized to distort the substrate, thereby decreasing the activation barrier. Although a rigid enzyme is not a necessary requirement, it is difficult to envision the protein distorting the substrate if the active site were itself more easily distorted. The rigid active site so strongly evokes the lock and key that we are led to search for a further extension of this mechanism to the explanation of enzyme catalysis, and in particular to the phenomenon of rate enhancement specificity.

A reacting molecule may undergo a volume change as it reaches the transition state. When the reaction occurs in the gas phase, the work associated with this volume difference is negligible. In a liquid medium molecular densities are relatively high, and the work due to the volume change can no longer be ignored. For example, in a reaction involving molecular expansion part of the energy required to reach the transition state is used for enlarging the liquid cavity within which the reacting molecule is found. The magnitude of this expansive work depends on the internal pressure of the liquid, the free energy increasing with the cohesive interaction between the liquid molecules. This effect of the liquid medium on reaction rates is well recognized and has been discussed previously [103]. To analyze the medium effect more closely we may divide the overall activation process into three arbitrary steps; vaporization of the reference state molecule, transformation of the gaseous reactant to its activated structure, and condensation of the activated molecule into the liquid. Conversion of the gaseous molecule to its transition state in the second step requires an intrinsic activation free energy, which arises from the various internal transitions of the molecule. This contribution to the overall activation free energy is not dependent on the nature of the liquid environment, since it has been assigned to the gas phase reaction. The effect of the medium is entirely reflected in the vaporization of the reference state and the condensation of the transition state molecules. We, therefore, define:

$$\Delta G^{\ddagger} = \Delta G^{\ddagger}_{int} + \Delta G^{\ddagger}_{med}$$
$$\Delta G^{\ddagger}_{med} = \Delta G_2^{\ddagger v} - \Delta G_2^{v}$$

(104)

where $\Delta G^{\ddagger}_{int}$, and $\Delta G^{\ddagger}_{med}$ are the medium independent, and dependent contributions to the overall activation free energy, $\Delta G^{\ddagger}$; ΔG_2^{v} and $\Delta G_2^{\ddagger v}$ are the vaporization free energies of the reference and transition state molecules respectively. The formalism of equation 104 applies regardless of the medium under consideration so that the reaction of a substrate at an enzyme active site may be handled in the same manner. To explain enzyme catalysis we need to account for the difference between the $\Delta G^{\ddagger}$ of the catalyzed and uncatalyzed reactions. Since $\Delta G^{\ddagger}_{int}$ is independent of the reaction environment, it will be identical for both reaction systems, and rate differences will derive solely from $\Delta G^{\ddagger}_{med}$. This term accounts for all interactions, including catalytic groups, provided by the environment. Proximity effects will, however, be specifically excluded by use of the rational thermodynamic convention.

Our primary concern is that interaction with the medium that arises from the volume change of the reacting molecule as it passes to the transition state. This will be but one of the many factors influencing $\Delta G^{\ddagger}_{med}$, and thus the reaction rate, but as a first approximation this, the hard volume component, may be isolated and treated separately. By reducing the problem to one of hard volumes, the need to account for activation enthalpies is dropped, as has been previously discussed. We are left with the calculation of that part of the vaporization entropy due to the geometrical constraints imposed by the medium on the hard volume representation of the molecule in its reference and transition states. A geometric problem, which does not require knowledge of intermolecular potentials, may in theory be solved exactly. However, the complete geometries of reference and transition state complexes formed between any enzyme and its substrate are not known. Moreover, the equations of state so far developed for fluids of hard particles apply only to spheres or simple distortions of spheres [116, 118, 119, 131]. Thus in order to solve even this simplest form of the activation energy problem we must go to a crude model built around the assumption of spherical symmetry. We start with a hypothetical unimolecular reaction of a hard sphere in which the transition state particle is also spherical but has a different diameter. The reaction occurs either in a hard sphere liquid characterized by the number density and collision diameter of water, or in an undeformable spherical cavity that represents an enzyme's active site.

The vaporization entropy of a hard sphere solute from a hard sphere fluid is given by equation 83. Thus, after assigning the diameters of reference and transition state particles, σ_2 and $\sigma_2^\ddagger$, $\Delta S_{\text{med}}^\ddagger(\text{liq})$ may be calculated. Rather than assign these diameters, for computational convenience entropy changes are calculated in terms of the differential $\partial \Delta S_{2}^v / \partial \sigma_2$. This differential is related to the total entropy change by:

$$\Delta S_{\text{med}}^\ddagger(\text{liq}) = \int_{\sigma_2}^{\sigma_2^\ddagger} \frac{\partial \Delta S_2^v(\text{liq})}{\partial \sigma_2} \, d\sigma_2. \tag{105}$$

Provided the reacting sphere is always smaller than the spherical cavity representing the active site, the vaporization entropy may be calculated with equation 4, and

$$\Delta S_{\text{med}}^\ddagger(\text{cav}) = R \ln \frac{v^{f\ddagger}}{v^f} \tag{106}$$

with v^f and $v^{f\ddagger}$ the free volumes of the sphere in the reference and transition states respectively. The free volume of a sphere within a spherical cavity is easily seen to be:

$$v^f = \frac{\pi}{6} (\sigma_c - \sigma)^3 \tag{107}$$

so that:

$$\frac{\partial \Delta S_2(\text{cav})}{\partial \sigma_2} = \frac{3R}{\sigma_2 - \sigma_c} \tag{108}$$

where σ_c is the cavity diameter and σ_2 is the diameter of the reference state particle. Rate enhancement or inhibition due to the hard volume interactions within the cavity as opposed to the liquid medium is given by:

$$\frac{k(\text{cav})}{k(\text{liq})} = \exp\left[\frac{1}{R}\int_{\sigma_2}^{\sigma_2^{\ddagger}}\frac{\partial\,\Delta\Delta S_2^{v}}{\partial\sigma_2}\,d\sigma_2\right]$$

$$\frac{k(\text{cav})}{k(\text{liq})} = \exp\left\{\frac{1}{R}\int_{\sigma_2}^{\sigma_2^{\ddagger}}\left[\frac{3R}{\sigma_c-\sigma_2}-\frac{\partial\,\Delta S_2^{v}(\text{liq})}{\partial\sigma_2}\right]d\sigma_2\right\}. \tag{109}$$

Although equation 109 may be solved by the insertion of equation 83, and thus explicitly evaluated, we have chosen to leave it in the implicit form and to calculate the differential of equation 83 with a digital computer. The quantity $\partial\,\Delta\Delta S_2^{v}/\partial\sigma_2$ was then computed for many possible combinations of cavity and sphere sizes. Only one set of results is presented in Figure 10A; qualitatively similar results were obtained using other cavity diameters.

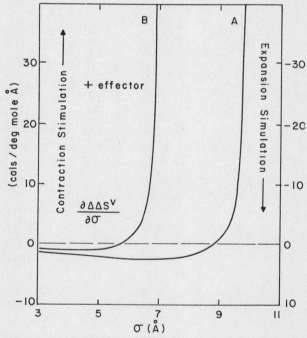

Figure 10   Hard volume contribution to activation entropies-unimolecular reaction. A. Difference between reactions in a hard cavity and hard sphere fluid. B. Difference between reactions in a hard cavity and hard sphere fluid in the presence of an inert sphere with diameter 3 Å. The calculations were performed as described in the text. The hard sphere fluid was assigned the number and collision diameter of water at 25°C. The cavity diameter is 10 Å. The results are presented in terms of the differential $\partial\Delta\Delta S_2^{v}/\partial\sigma_2$. The absolute rate enhancement for any set of diameters σ_2 and $\sigma_2^{\ddagger}$ is the area under the curve (equation 109).

For the hypothetical unimolecular reaction that involves a contraction to the transition state (Figure 10, left axis), the hard volume interaction between a small reacting sphere and the cavity results in a rate slower relative to that in the liquid phase; that is, negative catalysis is predicted. Because the free volume available to the small sphere is less in the liquid than in the cavity, contraction in the liquid releases more translational entropy; restated, the density is sufficiently greater in the liquid than in the cavity so as to promote contraction more effectively. As the sphere size approaches the cavity size the free volume in the cavity decreases sharply, and positive catalysis is observed. The results of Figure 10A apply to a model system that is only a crude representation of a real reaction. Thus the calculated magnitudes for the hard volume contribution to the activation entropies cannot be applied directly to enzyme catalysis. These results are qualitatively significant, however. The hard volume effect devolves upon the geometric interaction of the molecule with its environment. The introduction of more complex geometries as better molecular mimics will result in refinement of the calculations, but will not alter the conclusion that a reaction within a rigid cavity can occur with enhancement or inhibition depending on the geometries of cavity and reactant. Of special interest is the sensitive dependence of the hard volume entropy contribution on the size of the reacting sphere. This sensitivity (seen in Fig. 10A), as the sphere and cavity diameters become similar, indicates that a rigid cavity can in theory impose a specificity of rate enhancement on a reaction by providing the proper size. The same conclusions are obtained for a reaction involving expansion (Figure 10, right axis), because the results differ only by inclusion of a negative sign.

The hard sphere model used to calculate the results of Figure 10 is a lock-and-key mechanism in which specific electrostatic interactions have been ignored. In this one case we conclude that a rigid cavity can impose a specificity of rate enhancement and that no additional mechanism such as induced fit is required. Yet it is generally held that a lock-and-key mechanism cannot account for rate enhancement specificities. An important argument leading to this position is based on the observed lack of correlation for certain enzymes between maximum enzymatic rates and binding affinities for homologous compounds [130]. In at least one case, for example, turnover numbers are found to decrease, as affinities simultaneously increase over a series of compounds. (It should be noted that the hard sphere reaction with contraction displays the same dichotomy.) There is, however, a rigorous thermodynamic argument by which we can establish that such a lack of correlation does not require abandonment of the lock-and-key mechanism. The formalism expressed in equation 104 can be extended for this purpose. The catalytic ability of an enzyme is defined as the difference in activation free energies for a reaction occurring in water and at the enzyme's active site:

$$\Delta\Delta G^{\ddagger} = \Delta G^{\ddagger}(\text{enz}) - \Delta G^{\ddagger}(\text{H}_2\text{O}). \tag{110}$$

Because the intrinsic activation free energy $\Delta G^{\ddagger}_{\text{int}}$ is independent of the reaction medium, equation 110 becomes:

$$\Delta\Delta G^{\ddagger} = [\Delta G^{\ddagger v}(\text{enz}) - \Delta G^{\ddagger v}(\text{H}_2\text{O})] - [\Delta G_2{}^v(\text{enz}) - \Delta G_2{}^v(\text{H}_2\text{O})]. \tag{111}$$

But the difference in vaporization free energies of a solute from two different solvents is the free energy of transfer between the solvents, and the expressions in the brackets of equation 111 are the binding free energies of the reference and transition state molecules with the enzyme. We have arrived at the intuitively acceptable conclusion that, formally, enzyme catalysis means stronger binding of the transition state molecule. Equation 111 further states that catalytic ability depends on the binding specificity of the enzyme towards *both* reference and transition state molecules. For this reason we cannot argue that lack of correlation between catalytic rate and substrate affinity rules out a particular reaction mechanism. If we are willing to entertain the conjecture that a lock-and-key interaction can provide for substrate binding specificity, then we should logically be willing to accept the same mechanism as an explanation of rate enhancement specificity.

The calculations for the unimolecular reaction can be extended to a bimolecular case. Once again the transition state is taken as a sphere, because of the limitations of equation 83. Defining the diameters of the reacting spheres as σ_2 and σ_3, the diameter of the transition state particle resulting from their fusion is given by:

$$\sigma_{23} = (\mu_2\sigma_2{}^3 + \mu_3\sigma_3{}^3)^{1/3} \tag{112}$$

where μ_2 and μ_3 are relative densities that may be used to account for contractions or expansions. For the reaction in a fluid of hard spheres, the hard volume contribution to the activation entropy can be calculated with equation 83, as in the unimolecular case, after assigning values to the four parameters in the righthand side of equation 112. The free volume of an unreacted sphere contained inside a cavity together with a second sphere is given by the relationships of equation 113. The full derivation of these equations is given in a manuscript currently being prepared.

$$v_3{}^f = \frac{2\pi}{(\sigma_c - \sigma_2)^3 - (2\sigma_3 + \sigma_2 - \sigma_3)^3}$$
$$\times \left[\tfrac{3}{8}(a^2 - b^2)R^2 + \tfrac{2}{3}(b^3 - a^3)R^3 + \tfrac{3}{8}(b^2 + a^2)R^4 - \frac{R^6}{24}\right]_{R=(2\sigma_3+\sigma_2-\sigma_c)/2}^{R=(\sigma_c-\sigma_2)/2} \tag{113a}$$

when

$$\sigma_2 + 2\sigma_3 > \sigma_c > \sigma_2 + \sigma_3$$

$$v_3{}^f = \frac{2\pi}{(\sigma_c - \sigma_2)^3}$$

$$\times \left[\tfrac{3}{8}(a^2 - b^2)R^2 + \tfrac{2}{3}(b^3 - a^3)R^3 + \tfrac{3}{8}(a^2 + b^2)R^4 - \frac{R^6}{24} \right]_{R=(\sigma_c-\sigma_2-2\sigma_3)/2}^{R=(\sigma_c-\sigma_2)/2}$$

$$+ \frac{4}{3}\pi(b^3 - a^3)\left[1 - \frac{\sigma_3}{\sigma_2 - \sigma_3} \right] \quad (113b)$$

when

$$\sigma_2 + 2\sigma_3 \leqslant \sigma_c > \sigma_2 + \sigma_3.$$

R is the distance between the centers of the two spheres contained within the cavity of diameter σ_c. The parameters a and b are defined by:

$$a = (\sigma_2 + \sigma_3)/2$$
$$b = (\sigma_e - \sigma_3)/2. \quad (114)$$

The free volume, $v_3{}^f$, of the sphere with diameter σ_3 is given in equation 113; the free volume, $v_2{}^f$, of the second particle is obtained with the same equation by interchanging the subscripts. The free volume of the transition state sphere is calculated with equation 107.

The activation entropies for the bimolecular reaction in a hard sphere liquid or in a spherical cavity may be calculated with equations 83, 106, 112, and 113. The comparison of these entropies is a measure of the catalytic ability of the cavity. Examples of the results which can be obtained are presented in Figure 11. As would be expected the calculated curves yield conclusions similar to those discussed for the unimolecular case; the model displays both negative and positive catalysis, and rate enhancement specificity sensitive to the relative diameters of the cavity and the reacting spheres. Although no new conclusions can be drawn, this model can be modified to a consideration of the effect an inert particle would have on a reaction occurring in a rigid cavity. The addition of a second inert particle into a liquid solution would negligibly effect the rate of a reaction, provided the solution is sufficiently dilute. Within a rigid cavity an inert particle would reduce the free volume available to the reacting particle, and should, therefore, alter the hard volume activation entropy. This entropy term remains unchanged for a unimolecular reaction in a hard sphere fluid, but in the cavity the free volume is given by equation 113, not equation 107. With this substitution, and using the same conditions as those for the unimolecular reaction in the absence of the inert particle, the curve of Figure 10B is generated. The results for the two unimolecular reactions are qualitatively similar except for

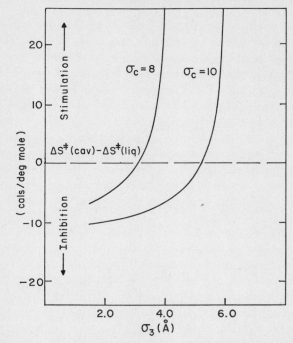

Figure 11   Hard volume contribution to activation entropies-bimolecular reaction. The calculations were performed as described in the text with the hard sphere liquid parameters described in Figure 10. The density parameters μ_2 and μ_3 were set to unity, and σ_2 was held constant at 4 Å. The two curves were calculated for cavities of 8 and 10 Å.

the effect of the inert particle in shifting the curve towards lower sphere diameters. Thus the inert particle either stimulates or inhibits the unimolecular reaction simply by occupying space within the rigid cavity, and the hard sphere model is found to include the possibility of a noncompetitive effector.

The model on which the calculations of this section are based is extremely crude. First, electrostatic interactions have been ignored, even though these strongly influence the reactions of interest to biochemists. As pointed out previously, however, it is permissible to isolate and separately consider the steric contribution to a reaction, because this term can be approximated as an independent factor in the overall process. The calculated results are not meant to mimic any real process, but are meant to indicate how hard volume interactions might impose their own requirements on a reaction. Second, although the experimental data previously cited tend to support the concept of a rigid active site, an undeformable active site would be clearly impossible. A deformable cavity has not been included in the model introducing an error the magnitude of which becomes extremely large in the limiting case of identical sphere and cavity sizes. Finally, the geometry of the model is clearly incorrect. Substrates and active sites are not spherical; isotropic changes to the transition state would not be expected for any reaction; and no shape factors have been introduced for the bimolecular case. In spite of all these shortcomings, the conclusions drawn from this model are sufficiently interesting to warrant refinements in the calculations and experimental test. To this reviewer's knowledge there is currently only one experimental result that may be applicable to this discussion. The trypsin catalyzed hydrolysis of acetylglycine ethyl ester is stimulated by methyl- and ethylamine, and by methyl- and ethylguanidine, while the longer chain alkyl compounds inhibit [138]. It has been proposed that these small basic solutes occupy the cavity that normally holds the lysine and arginine side chains, and thus induce a conformational change leading to a better protein structure for catalysis. In other words, the stimulation has been interpreted in terms of an induced fit mechanism. It is also feasible to propose that these compounds stimulate the ester hydrolysis simply by occupying the side chain pocket and changing the activation entropy by the hard volume mechanism discussed in this section. This interpretation is attractive in light of the observation that the ethylamine and methylguanidine stimulations arise primarily from entropy increases, the enthalpy contribution being negligible [138].

Recently Page and Jencks [139] have proposed an entropic argument for enzymatic rate enhancement. Their proposal is complementary to the model discussed here, but is not similar. A protein imposes restrictions on a bimolecular reaction due to a concentration effect at the active site and to alteration of rotational movement. Page and Jencks have argued that these restrictions result in entropy changes sufficiently large so as to obviate the requirement for conformational changes by the enzyme. Neither of these effects has been introduced into the hard volume model. The first was specifically excluded by the use of the unitary thermodynamic scale; the second was omitted since the reacting particles were taken as hard spheres.

5 CONCLUSION

Since its proposal, the concept of the apolar bond has been associated closely with the theory of water structure stabilization and with the small molecule model of liquid-liquid partition. The theory has provided the imagery needed for psychological acceptance, and the model has supplied the thermodynamic parameters required for predictive purposes. As a result the apolar bond has been afforded a physically concrete picture, but it has not been exactly defined. A model may serve to suggest the mechanism of a more complicated process or to yield experimental results with which to approximate the data not readily extracted from the more complicated case. Although the liquid-liquid partition model has been immensely important for its emphasis on the significance of solvent interactions with apolar residues, it has not been applied quantitatively with success. It is for this reason that the foundations upon which the apolar bond has been constructed are reevaluated in this review. I began with the hope that a consideration of the experimental and theoretical results obtained after the original formulation of the apolar bond would provide the framework for a quantitative application to proteins. I finish with the belief that this is not possible.

There is evidence which suggests that structural stabilization of solvent does not contribute significantly to the solution of apolar molecules in water. Furthermore, were such a process significant, it would still not be felt in the free energy of solution; that is, the low solubility of apolar solutes in water cannot be due to the stabilization of lattice-like water, and this process cannot be invoked as the force driving apolar residues into the protein interior. There is also reason to suspect that at least one of the premises upon which the liquid-liquid partition model is based, the oil-drop nature of the protein interior, may be incorrect. Continued use of this model could, therefore, hinder rather than aid further understanding of protein chain unfolding. To abandon both the theory of the solution mechanism and the thermodynamic model without simultaneously losing the concept would require a new description of apolar bonding in proteins. There are approaches by which a redescription could be attempted. For example, a different small molecule model may be adapted, but this procedure leaves us with the task not yet completed for the old, determining the validity of this new model system. It may be possible to use one aspect of the current theory as a point of departure for the new. An important emphasis of the apolar residue is the governing role played by the interaction between the apolar protein residues and the aqueous environment. This importance is reflected in the introduction of the adjectives hydrophobic, lyophobic, and solvophobic. But building upon this one emphasis may not be fruitful. It is, for example, well accepted that ionic molecules transfer readily from an apolar to an aqueous medium driven by hydration of the charged species. There does not, however, appear to be any need to formalize this process with the name of hydrophilic bonding, because it is already understood in terms of presently formulated concepts.

It is my opinion that the apolar bond cannot be reformulated so as to satisfy both our intuitive feeling that apolar side chains dislike water, and the experimental observations that are in contradiction to our present conception. It would be better perhaps to realize that when the trees are cut down, the grove cannot be saved, and that it may prove more fruitful to explain the experimental observation of apolar cores in proteins in terms of the same chemical principles used in the explanation of other molecular interactions. A reformulation on this basis has not been attempted here. One approach towards this task may be a further application of the Van der Waals approximation. The successful use of the scaled particle theory to describe the solution of apolar gases in liquids suggests that repulsive, hard volume interactions may be generally important for many solutions and encourages the attempt to expand the theory to systems that clearly cannot be represented with spherical symmetry. The hard volume approximation has, in fact, already been successfully employed as a first approximation in the construction of Ramachandran maps [81]. The possibility that steric restraints impose the major restrictions on the secondary structure of polypeptide chains raises the further possibility that steric interactions may be important in determining the tertiary structure as well. It remains to be determined whether such an approach will yield new and important insights for the question of what makes a protein fold as it does.

Finally, I would like to acknowledge my gratitude to the many people who were willing to pay attention either to me or to this manuscript, and then to offer helpful comments. The free time on the IBM 360 given by the Ohio State Computer Center is gratefully acknowledged. And I give profound thanks to my secretary, B. Cassity, who retyped the countless revisions with only the faintest indications of impatience.

REFERENCES

1. W. Kauzmann, *Adv. Prot. Chem.*, **14,** (1959), p. 1.
2. F. M. Richards, *Ann. Rev. Biochem.*, **32,** 269 (1963).
3. W. P. Jencks, *Catalysis in Chemistry and Enzymology*, McGraw-Hill, New York, 1969.
4. J. F. Brandts in *Structure and Stability of Biological Macromolecules*, S. N. Timasheff and G. D. Fasman (Eds.), Marcel Dekker, New York, 1969 p. 213.
5. G. Nemethy, *Angew. Chem.*, **6,** 195 (1967).
6. I. M. Klotz, Brookhaven Symp. Biol., **13,** 25 (1960).
7. D. D. Eley, *Trans. Farad. Soc.*, **35,** 1281 (1939).
8. J. A. V. Butler, *Trans. Farad. Soc.*, **33,** 229 (1937).
9. R. P. Bell, *Trans. Farad. Soc.*, **33,** 496 (1937).
10. J. E. Leffler and E. Grunwald, *Rates and Equilibria of Organic Reactions*, Wiley, New York, 1963, p. 128.
11. H. S. Frank and M. W. Evans, *J. Chem. Phys.*, **13,** 507 (1945).
12. G. C. Kresheck, H. Schneider, and H. A. Scheraga, *J. Phys. Chem.*, **69,** 3132 (1965).
13. R. Lumry and R. Biltonen in *Structure and Stability of Biological Macromolecules*, S. N. Timasheff and G. D. Fasman (Eds.) Marcel Dekker, New York, 1969, p. 65.
14. E. B. Smith and J. Walkley, *J. Phys. Chem.*, **66,** 597 (1962).
15. K. W. Miller and J. H. Hildedrand, *J. Am. Chem. Soc.*, **90,** 3001 (1968).
16. R. A. Pierotti, *J. Phys. Chem.*, **69,** 281 (1965).
17. W. L. Masterton, *J. Chem. Phys.*, **22,** 1830 (1954).
18. H. S. Frank, *Fed. Proc.*, **24,** S-1 (1965).

19. D. Eisenberg and W. Kauzmann, *The Structure and Properties of Water*, Oxford University Press, New York, 1969.
20. J. L. Kavanau, *Water and Solute-Water Interactions*, Holden-Day, San Francisco, 1964.
21. R. W. Gurney, *Ionic Processes in Solution*, McGraw-Hill, New York, 1953.
22. J. H. Hildebrand and R. L. Scott, *The Solubility of Nonelectrolytes*, 3rd ed., Dover Publications, New York, 1964.
23. T. L. Hill, *An Introduction to Statistical Thermodynamics*, Addison-Wesley, London, 1960, p. 288.
24. H. S. Frank and W. Y. Wen, *Disc. Farad. Soc.*, **24**, 133 (1957).
25. H. S. Frank, *Proc. Roy. Soc., Lond.*, **A247**, 481 (1958).
26. H. S. Frank and A. S. Quist, *J. Chem. Phys.*, **34**, 604 (1961).
27. H. S. Frank and F. Franks, *J. Chem. Phys.*, **48**, 4746 (1968).
28. R. M. Barrer in *Non-Stoichiometric Compounds*, L. Mandelcorn (Ed.), Academic Press, New York, 1964, p. 309.
29. G. A. Jeffrey and R. K. McMullan, *Prog. Inorg. Chem.*, **8**, 43 (1967).
30. G. A. Jeffrey, *Acc. Chem. Res.*, **2**, 344 (1969).
31. M. von Stackelberg and W. Meinhold, *Z. Electrochem.*, **58**, 40 (1954).
32. J. G. Waller, *Nature*, **186**, 429 (1960).
33. P. T. Beurskens and G. A. Jeffrey, *J. Chem. Phys.*, **40**, 2800 (1964).
34. J. H. van der Waals and J. C. Platteeuw, *Adv. Chem. Phys.*, **2**, 1 (1959).
35. W. C. Child, *J. Phys. Chem.*, **68**, 1834 (1964).
36. W. C. Child, *Quart. Rev.*, **18**, 321 (1964).
37. R. M. Barrer and A. V. J. Edge, *Proc. Roy. Soc.*, **A300**, 1 (1967).
38. R. A. Pierotti, *J. Phys. Chem.* **67**, 1840 (1963).
39. R. Battino, H. L. Clever, *Chem. Rev.*, **66**, 395 (1966).
40. L. W. Reeves and J. H. Hildebrand, *J. Am. Chem. Soc.*, **79**, 1313 (1957).
41. L. Stryer, *Ann. Rev. Biochem.*, **37**, 25 (1968).
42. I. M. Klotz, *Arch. Biochem. Biophys.*, **138**, 704 (1970).
43. G. Nemethy and H. A. Scheraga, *J. Chem. Phys.*, **36**, 3382 (1962).
44. G. Nemethy and H. A. Scheraga, *J. Chem. Phys.*, **36**, 3401 (1962).
45. M. Rigby, *Quart. Rev.*, **24**, 416 (1970).
46. H. Reiss, *Adv. Chem. Phys.*, **9** (1965).
47. J. L. Lebowitz, E. Helfand, and E. Praestgaard, *J. Chem. Phys.*, **43**, 774 (1965).
48. J. K. Percus and G. J. Yevick, *Phys. Rev.*, **110**, 1 (1958).
49. J. O. Hirschfelder, C. F. Curtiss, and R. B. Bird, *Molecular Theory of Gases and Liquids*, Wiley, New York, 1954.
50. S. J. Yosim and B. B. Owens, *J. Chem. Phys.*, **39**, 2222 (1963).
51. J. B. Scarborough, *Numerical Mathematical Analysis*, 2nd ed., The Johns Hopkins Press, Baltimore, 1950, p. 192.
52. S. W. Mayer, *J. Phys. Chem.*, **67**, 2160 (1963).
53. F. Steckel and S. Szapiro, *Trans. Farad. Soc.*, **59**, 331 (1963).
54. *International Critical Tables* **3**, 28 (1928).
55. R. A. Pierotti, *J. Phys. Chem.*, **67**, 1840 (1963).
56. S. J. Yosim, *J. Chem. Phys.*, **43**, 286 (1965).
57. T. Boublik and G. C. Benson, *J. Phys. Chem.*, **74**, 904 (1970).
58. J. A. Barker and D. Henderson, *J. Chem. Phys.*, **47**, 4714 (1967).
59. J. A. Barker and D. Henderson, *J. Chem. Ed.*, **45**, 2 (1968).
60. D. Henderson and J. A. Barker, *J. Chem. Phys.*, **49**, 3377 (1968).
61. H. H. Uhlig, *J. Phys. Chem.*, **41**, 1215 (1937).
62. J. H. Saylor and R. Battino, *J. Phys. Chem.*, **62**, 1334 (1958).

63. E. McLaughlin, *J. Chem. Phys.*, **50,** 1254 (1969).
64. J. H. Wang, C. V. Robinson, and I. S. Edelman, *J. Am. Chem. Soc.*, **75,** 466 (1953).
65. H. Watts, B. J. Alder, and J. H. Hildebrand, *J. Chem. Phys.*, **23,** 659 (1955).
66. M. Ross and J. H. Hildebrand, *J. Chem. Phys.*, **40,** 2397 (1964).
67. E. W. Haycock, B. J. Alder, and J. H. Hildebrand, *J. Chem. Phys.*, **21,** 1601 (1953).
68. D. N. Glew, H. D. Mak, and N. S. Rath in *Hydrogen-Bonded Solvent Systems*, A. K. Covington and P. Jones (Eds.), Taylor and Francis, London, 1968, p. 195.
69. P. H. von Hippel and T. Schleich in *Structure and Stability of Biological Macromolecules*, S. N. Timasheff and G. D. Fasman (Eds.), Marcel Dekker, New York, 1969, p.417.
70. B. J. Alder and T. E. Wainwright, *J. Chem. Phys.*, **33,** 1439 (1960).
71. A. van Cleef and G. A. M. Diepen, *Rec. Trav. Chim.*, **84,** 1085 (1965).
72. A. Ben-Naim, *J. Phys. Chem.*, **69,** 1922 (1965).
73. A. Ben-Naim, *J. Phys. Chem.*, **71,** 4002 (1967).
74. A. Ben-Naim and S. Baer, *Trans. Farad. Soc.*, **60,** 1736 (1964).
75. D. B. Wetlaufer, S. K. Malik, L. Stoller, and R. L. Coffin, *J. Am. Chem. Soc.*, **86,** 508 (1964).
76. F. A. Long and W. F. McDevit, *Chem. Rev.*, **51,** 119 (1952).
77. A. Holtzer and M. F. Emerson, *J. Phys. Chem.*, **73,** 26 (1969).
78. W. F. McDevit and F. A. Long, *J. Am. Chem. Soc.*, **74,** 1773 (1952).
79. S. K. Shoor and K. E. Gubbins, *J. Phys. Chem.*, **73,** 498 (1969).
80. W. L. Masterton and T. P. Lee, *J. Phys. Chem.*, **74,** 1776 (1970).
81. C. M. Venkatachalam and G. N. Ramachandran, *Ann. Rev. Biochem.*, **38,** 45 (1969).
82. G. G. Hammes and P. R. Schimmel, *J. Am. Chem. Soc.*, **89,** 442 (1967).
83. E. P. K. Hade and C. Tanford, *J. Am. Chem. Soc.*, **89,** 5034 (1967).
84. I. D. Kuntz, T. S. Brassfield, G. D. Law, and G. V. Purcell, *Science*, **163,** 1329 (1969).
85. R. Jaenicke and M. A. Lauffer, *Biochemistry*, **8,** 3077 (1969).
86. H. B. Bull and K. Breese, *Arch. Biochem. Biophys.*, **139,** 93 (1970).
87. B. Blicharska, Z. Florkowski, J. W. Hennel, G. Held, and F. Noack, *Biochim. Biophys. Acta*, **207,** 381 (1970).
88. M. H. Klapper, *Biochim. Biophys. Acta* **229,** 557 (1971).
89. J. T. Edward, *J. Chem. Ed.* **47,** 261 (1970).
90. R. A. Harte and J. A. Rupley, *J. Biol. Chem.* **243,** 1663 (1968).
91. J. F. Brandts, R. J. Oliveira, and C. Westort, *Biochemistry* **9,** 1038 (1970).
92. I. M. Klotz, N. R. Langerman, and D. W. Darnall, *Ann. Rev. Biochem.*, **39,** 25 (1970).
93. G. J. Ewing and S. Maestas, *J. Phys. Chem.*, **74,** 2341 (1970).
94. A. Wishnia, *Biochemistry*, **8,** 5070 (1969).
95. H. S. Frank, *J. Chem. Phys.*, **13,** 493 (1945).
96. B. P. Schoenborn, H. C. Watson, and J. C. Kendrew, *Nature*, **207,** 28 (1965).
97. B. P. Schoenborn, *J. Mol. Biol.* **45,** 297 (1969).
98. R. G. Shulman, J. Peisach, and B. J. Wyluda, *J. Mol. Biol.*, **48,** 517 (1970).
99. W. N. Lipscomb, J. A. Hartsuck, F. A. Quiocho, and G. N. Reeke, *Proc. Nat. Acad. Sci. (U.S.)* **64,** 28 (1969).
100. S. T. Freer, J. Kraut, J. D. Robertus, H. T. Wright, and N. H. Xuong, *Biochemistry*, **9,** 1997 (1970).
101. H. C. Watson and B. Chance in *Hemes and Hemoproteins*, B. Chance, R. W. Estabrook, and T. Yonetani (Eds.), Academic Press, New York, 1966, p. 149.
102. H. Muirhead, J. M. Cox, L. Mazzarella, and M. F. Perutz, *J. Mol. Biol.*, **28,** 117 (1967).
103. S. Glasstone, K. J. Laidler, and H. Eyring, *The Theory of Rate Processes*, McGraw-Hill, New York, 1941, p. 413.
104. G. Lasher, *J. Chem. Phys.*, **53,** 4141 (1970).

105. J. T. Ashton, R. A. Dawe, K. W. Miller, E. B. Smith, and B. J. Stickings, *J. Chem. Soc.*, 1793 (1968).
106. D. N. Glew, E. A. Moelwyn-Hughes, *Disc. Farad. Soc.*, **15**, 150 (1953).
107. J. A. Rupley in *Structure and Stability of Biological Macromolecules*, S. N. Timasheff and G. D. Fasman (Eds.), Marcel Dekker, New York, 1969, p. 291.
108. A. J. Hymes, D. A. Robinson, and W. J. Canady, *J. Biol. Chem.*, **240**, 134 (1965).
109. G. Royer and W. J. Canady, *Arch. Biochem. Biophys.*, **124**, 530 (1968).
110. F. Helmer, K. Kiehs, and C. Hansch, *Biochemistry*, **7**, 2859 (1968).
111. A. N. Glazer, *Proc. Nat. Acad. Sci.*, **65**, 1057 (1970).
112. P. J. Flory, *Statistical Mechanics of Chain Molecules*, Interscience, New York, 1969, p. 302.
113. Y. Nozaki and C. Tanford, *J. Biol. Chem.*, **246**, 2211 (1971).
114. D. M. Blow and T. A. Steitz, *Ann. Rev. Biochem.*, **39**, 63 (1970).
115. B. Lee and F. M. Richards, *J. Mol. Biol.*, **55**, 379 (1971).
116. D. M. Tully-Smith and H. Reiss, *J. Chem. Phys.* **53**, 4015 (1970).
117. M. Lucas and A. deTrobriand, *J. Phys. Chem.*, **75**, 1803 (1971).
118. M. A. Cotter and D. E. Martire, *J. Chem. Phys.*, **53**, 4500 (1970).
119. L. K. Runnels and C. Colvin, *J. Chem. Phys.*, **53**, 4219 (1970).
120. D. F. Evans and R. E. Richards, *Proc. Roy. Soc.*, **A223**, 238 (1954).
121. D. C. Phillips, personal communication, cited in reference 115.
122. G. N. Lewis and M. Randall, *Thermodynamics and the Free Energy of Chemical Substances*, McGraw-Hill, New York, 1923, p. 485.
123. H. W. Wyckoff, K. D. Hardman, N. M. Allewell, T. Inagami, L. N. Johnson, and F. M. Richards, *J. Biol. Chem.*, **242**, 3984 (1967).
124. R. A. Alden and C. S. Wright, J. Kraut, *Phil. Trans. Roy. Soc. Lond.*, **B257**, 119 (1970).
125. J. J. Birktoft, D. M. Blow, and R. Henderson, T. A. Steitz, *ibid*, p. 67.
126. B. G. Wolthers, J. Drenth, J. N. Jansonius, R. Koekoek, and H. M. Swen in *Structure-Function Relationships of Proteolytic Enzymes*, P. Desnuelle, H. Neurath, M. Ottesen (Eds.), Academic Press, New York, 1970, p. 272.
127. A. Arnone, C. J. Bier, F. A. Cotton, V. W. Day, E. E. Hazen, D. C. Richardson, J. S. Richardson, and A. Yonath, *J. Biol. Chem.*, **246**, 2302 (1971).
128. M. J. Adams, G. C. Ford, R. Koekoek, P. J. Lentz, A. McPherson, M. G. Rossmann, I. E. Smiley, R. W. Schevitz, and A. J. Wonacott, *Nature*, **227**, 1098 (1970).
129. I. E. Smiley, R. Koekoek, M. J. Adams, and M. G. Rossmann, *J. Mol. Biol.*, **55**, 467 (1971).
130. D. E. Koshland in *The Enzymes*, Vol. I, P. D. Boyer, H. Lardy, and K. Myrback (Eds.), Academic Press, New York, 1959, p. 305.
131. Y. D. Chen and W. A. Steele, *J. Chem. Phys.*, **54**, 703 (1971).
132. M. K. Tham and K. E. Gubbins, *J. Chem. Phys.*, **55**, 268 (1971).
133. E. Wilhelm and R. Battino, *J. Chem. Phys.*, **56**, 563 (1972).
134. R. Wildnauer and W. J. Canady, *Biochemistry*, **5**, 2885 (1966).
135. J. Greer, *J. Mol. Biol.*, **59**, 107 (1971).
136. P. M. Wassarman and P. J. Lentz, *J. Mol. Biol.*, **60**, 509 (1971).
137. D. E. Koshland and K. E. Neet, *Ann. Rev. Biochem.*, **37**, 359 (1968).
138. T. Inagami and S. S. York, *Biochemistry*, **7**, 4045 (1969).
139. M. I. Page and W. P. Jencks, *Proc. Nat. Acad. Sci.* (*U.S.A.*), **68**, 1678 (1971).

ON THE MECHANISMS
OF ACTION OF FOLIC
ACID COFACTORS

STEPHEN J. BENKOVIC

Department of Chemistry, The Pennsylvania State University,
University Park, Pennsylvania

WILSON P. BULLARD

Department of Chemistry, The Pennsylvania State University,
University Park, Pennsylvania

1 INTRODUCTION

In the past decade the role of 5,6,7,8-tetrahydrofolic acid-(H_4-folate) in the metabolism of single carbon units at varying levels of oxidation has been well established through the efforts of many investigators [1]. Although the major pathways and reactions involving these cofactors have been well described, studies concerned with purification and reaction characteristics of the specific enzymes involved have yielded, in general, only a limited amount of unambiguous and instructive physical and chemical data. As a consequence, the mechanisms of enzymic catalysis of many of these reactions are not well understood. Even the identity of certain proteins is still in question, because contaminating or overlapping activities remain unresolved. There are at this time, however, sufficient data from both the enzymic and model reactions, particularly on the oxidation level of formaldehyde and formate, to warrant discussion of the more probable catalytic mechanisms. The proposed mechanisms are unavoidably hypothetical and are formulated to emphasize mechanistic features that might be common to two or more of the individual transformations. In order to avoid excessive speculation, we have refrained from attempting a complete review of all the known reactions involving folic acid cofactors until warranted by the accumulation of evidence. However, two such areas are discussed briefly—methyl transfer and the oxidative-reductive interconversions of folic acid cofactors—to acquaint the reader with the present state of the science.

2 OXIDATIVE-REDUCTIVE INTERCONVERSIONS

For the successful utilization of exogenous folic acid, or of the 7,8-dihydrofolic acid formed in the biosynthetic pathway, the key enzyme is dihydrofolate reductase, which catalyzes the reduction of these species to the tetrahydro level. Blakley has summarized the properties of the enzyme isolated from various sources [1]. Since in most cases blockage of H_4-folate formation leads to the death of the cell because of the latter's inability to synthesize deoxythymidylic acid, one of the four nucleotide precursors to DNA, a great deal of effort has been expended to synthesize compounds that inhibit irreversibly and specifically the action of dihydrofolate reductase [2]. Such materials are exemplified by the isosteric analogs aminopterin 1b, and amethopterin 1c that stoichiometrically inhibit the enzyme owing to a tight binding constant ($>10^9$). The structure of H_4-folate 1a serves as a means for structural comparison.

Baker recently has synthesized a diaminopyrimidine alkylating agent 2 that is tissue specific *in vitro*. This response to the structure of 2 apparently is a consequence of subtle changes in the conformation of a hydrophobic

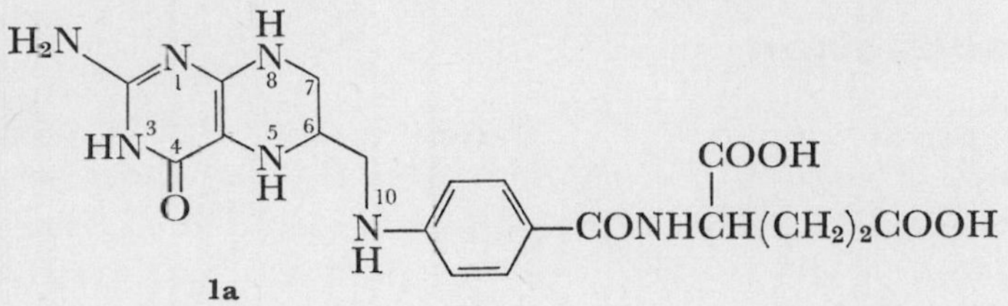

1a

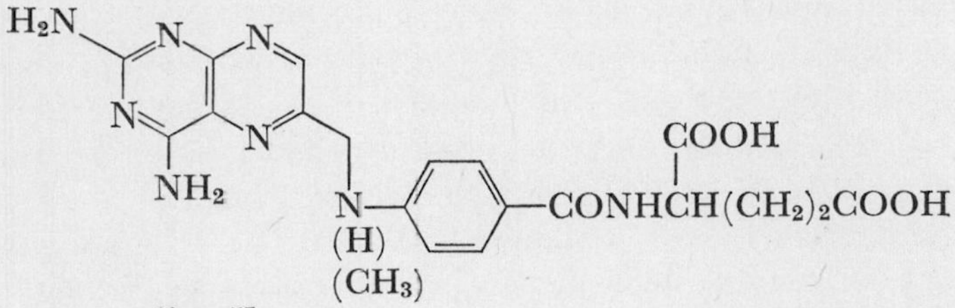

1b = H
1c = CH$_3$

binding site between the various reductases [3]. Transport problems, however, have deterred its successful employment *in vivo*. Although this approach has served to map the regions on both substrate and enzyme that participate in formation of the enzyme-cofactor complex, it reveals little about the mechanism of action of dihydrofolate reductase.

The borohydride reduction of folic acid offers a chemical model for the expected enzymic sequence of reactions [4]. The initial product of borohydride, as well as hydrosulfite reduction of folic acid, is the 7,8-dihydrofolic acid isomer that in turn is further and more rapidly reduced to the tetrahydro

$$\text{Folic acid} \xrightarrow{\text{H}_2} \text{7,8-H}_2\text{-folate} \xrightarrow{\text{H}_2} \text{H}_4\text{-folate} \tag{1}$$

level [5, 6]. Enzymic reduction of folic acid by NADPH in the presence of tritiated water does not lead to tritium incorporation into the 7,8-dihydro product, implying direct transfer of hydrogen from the reduced nicotinamide cofactor to C-7 of folic acid [7]. Conclusive nmr evidence now has been presented by Pastore, showing that hydrogen is transferred from the A side of NADPH to C-6 of 7,8-H$_2$-folate to complete the reduction to H$_4$-folate

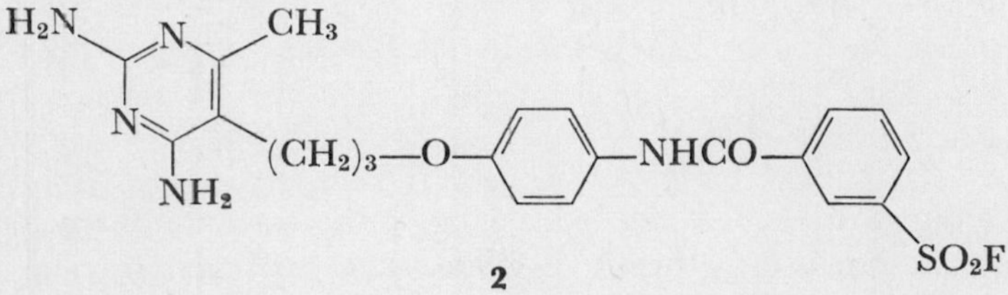

2

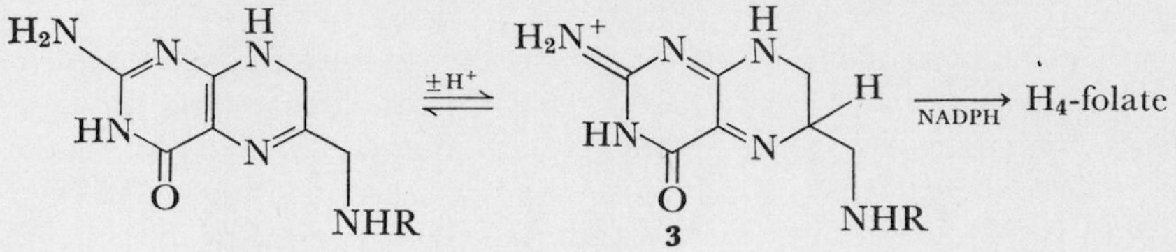

$$\xrightarrow{\pm H^+}$$ ⇌ $$\xrightarrow{NADPH} H_4\text{-folate} \qquad (2)$$

$$R = pC_6H_4\text{-CONHCH}(CO_2H)CH_2CH_2CO_2H$$

for the reductases isolated from *E. coli*, chicken liver, and mammalian tumor [8, 9]. The equilibrium greatly favors H_4-folate. There is no support for the hypothesis that the latter reduction proceeds via a prior intramolecular rearrangement of 7,8-H_2-folate that enables the incoming hydrogen to add at *C*-7 [10]. The addition of bisulfite ion to folic or dihydrofolic acid also follows a parallel course [11]. Although the stereochemical course for these hydrogen transfers is now partially elucidated, it still is not possible to reconstruct the absolute stereochemical reaction pathway. A major impediment is the unknown absolute configuration of H_4-folate itself. The question of the reaction mechanism involving the nicotinamide cofactors, likewise, remains open [12]. It appears doubtful, however, that the reactions on the dihydrofolic acid level involve prior rearrangement to a quinoid intermediate **3** in view of the results of Pastore demonstrating C—H rather than N—H bond formation in the reduction step. A similar quinoid species that spontaneously rearranges to 7,8-H_2-pteridine is produced in the phenylalanine hydroxylase reaction from 6,7-dimethyltetrahydropteridine [13–15]. Further aspects of this reaction have already been discussed in the preceding volume (see the chapter mentioned in reference 12).

A final oxidative-reductive sequence is considered, namely the interconversion of 5,10-methylene-H_4-folate and 5,10-methenyl-H_4-folate mediated by NADPH and the enzyme 5,10-methylene-H_4-folate dehydrogenase.

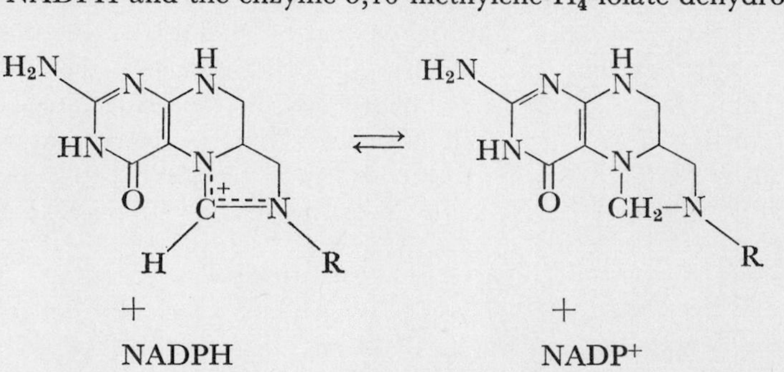

$$\rightleftharpoons \qquad (3)$$

+   +

NADPH   NADP+

The oxidized product is favored, owing to the concomitant nonenzymic hydrolysis of the 5,10-methenyl-H_4-folate. The hydrogen transfer process evidently is stereospecific because (1) the ethanol formed through coupled enzymic conversions commencing with [3]HCOOH and proceeding through serine is predominantly in the S configuration and (2) the dehydrogenase from yeast adds hydrogen to side A of the nicotinamide ring of NADP[+] [16, 17a]. Although one may surmise that hydrogen is added to the convex face of the cofactor, that is, from the same side as the hydrogen at C-6, this is not known. However, it is now known that a methylene bridge hydrogen is transferred directly to NADP[+] [17b]. In contrast, the further enzymic reduction of 5,10-methylene-H_4-folate to 5-methyl-H_4-folate catalyzed by 5,10-methylene-H_4-folate reductase from $E. coli$ requires flavin adenine dinucleotide as the immediate reductant so that labeling experiments involving the flavin are precluded by exchange with solvent [18, 19]. It is tempting to speculate that the species undergoing reduction in this case is an iminium cation formed at N-5, a key intermediate in the following section.

3 TRANSFER OF THE ONE-CARBON UNIT AT THE OXIDATION LEVEL OF FORMALDEHYDE

The formation of 5,10-methylene tetrahydrofolic acid from formaldehyde and the parent cofactor occurs rapidly in dilute solution in the absence of enzyme. The overall equilibrium lies far on the side of adduct ($K_{eq} = 3.2 \times 10^4\, M^{-1}$, pH 7.2), a fact that suggests that the structure of the condensation product involves a methylene bridge between nitrogens 5 and 10 rather than simply a N-hydroxymethyl derivative [20]. Additional support for this structure assignment is furnished by evidence that enzymic or borohydride reduction of 5,10-methenyl-H_4-folate produces this adduct [22, 23].

The mechanism of the condensation of formaldehyde with tetrahydrofolic acid has been extensively investigated by Kallen and Jencks [21]. In the absence of mercaptoethanol the reaction exhibits a bell-shaped pH-rate profile with a maximum at about pH 6.0 and inflection points at about pH 5.0 and 7.0. Because the latter pH does not correspond to a pK_a of either reactant, the pH-rate profile must be controlled by the kinetics of the reaction rather than the respective dissociation constants. Similar behavior has been encountered in a number of carbonyl addition reactions so that the kinetic interpretation of rate-determining attack by amine (presumably the more nucleophilic N-5) on formaldehyde at low pH changing to rate-limiting dehydration of the carbinolamine intermediate at alkaline pH has ample precedent [24]. Both steps are subject to general acid catalysis through the mechanisms illustrated by the transition states **4** and **5**.

$$\backslash_{\overset{\delta+}{HN^5}}\cdots CH_2\mathbin{\hspace{-0.3em}=\hspace{-0.3em}}O\cdots H\cdots \overset{\delta-}{A} \qquad \backslash_{\overset{\delta+}{N^5}}\overset{\overset{\textstyle H}{|}}{\underset{}{\cdots}}CH_2\cdots O\cdots H\cdots \overset{\delta-}{A}$$

$$\qquad\qquad\textbf{4}\qquad\qquad\qquad\qquad\qquad\textbf{5}$$

The same authors observed that the reaction of H_4-folate with formaldehyde is catalyzed by secondary amines such as morpholine and imidazole. Because the rate of the amine catalyzed reaction reaches a constant value at high concentrations of amine and increases with increasing acidity, the catalysis apparently involves condensation of the amine catalyst with formaldehyde to form a hydroxymethylamine. The latter then undergoes acid catalyzed dehydration to furnish a reactive cationic iminium species (4a

$$\backslash\!\!\!\diagup NH + CH_2O \rightleftharpoons \backslash\!\!\!\diagup N\!\!-\!\!CH_2OH \overset{\pm H^+}{\rightleftharpoons} \backslash\!\!\!\diagup \overset{+}{N}\!\!=\!\!CH_2 + H_2O \qquad (4a)$$

and 4b). The complete amine catalyzed reaction then would involve attack of the N-5 of H_4-folate on the iminium species followed by expulsion of the secondary amine and subsequent rapid cyclization.

$$\backslash\!\!\!\diagup \overset{+}{N}\!\!=\!\!CH_2 + N^5H \overset{\mp H^+}{\rightleftharpoons} \backslash\!\!\!\diagup N\!\!-\!\!CH_2\!\!-\!\!N^5 \overset{\pm H^+}{\rightleftharpoons}$$

$$\backslash\!\!\!\diagup NH + CH_2\!\!=\!\!\overset{+}{N^5} \overset{\mp H^+}{\rightleftharpoons} 5,10\text{-methylene-}H_4\text{-folate} \qquad (4b)$$

It was not possible to determine which of these latter steps is rate-determining.

In a study utilizing tetrahydroquinoxaline analogs, Benkovic et al. examined the condensation of formaldehyde with these models [25]. The design of these compounds, predicated on the desirability of simplifying experimental kinetic problems through the removal of several "nonessential" dissociable groups, retained, however, the molecular dissymmetry inherent in the parent cofactor, and, in the case of p-carbethoxy substitution, pK_a values for the participating nitrogen atoms closely approaching those for H_4-folate [26].

Two pK_a's were detected for **6,** 4.35, assigned to the quinoxaline ring nitrogens and -1.10, attributed to the exocyclic amino function. For comparison the two relevant pK_a's of H_4-folate are 4.82 and -1.25 for N-5 and N-10, respectively. Capitalizing on the observation that preequilibrium formation of an intermediate—tentatively identified as a carbinolamine—occurred between pH 4 and 8, Benkovic et al. [25] examined the conversion of this species—presumably formed at N-1—to the imidazolidine product. This reaction did not follow the anticipated pH-rate profile for simple acid-catalyzed dehydration of a carbinolamine intermediate but featured two regions dependent on hydronium ion concentration joined by a pH independent area (Fig. 1). The kinetic results were explained in terms of a steady-state species, the iminium cation **7,** the partitioning of which is subject to pH control according to the mechanism outlined (5).

It is noteworthy that this scheme features general base catalysis of attack by the exocyclic amino group on the iminium cation to yield the imidazolidine. Supporting evidence for this mechanism was sought by systematically increasing the basicity of N-10 through changes in the p-substituent for the series p-CH_3 > p-Cl > p-$COOC_2H_5$. One predicts a decrease in the magnitude of catalysis associated with cyclization and an increase in the rate of ring closure with increasing basicity at N-10. Both predictions were confirmed experimentally [27].

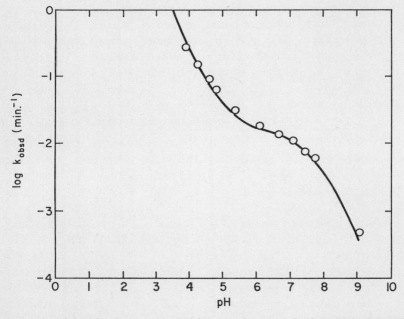

Figure 1   Plot of log k_{obsd} versus pH for ring closure of the formaldehyde adduct of **6.**

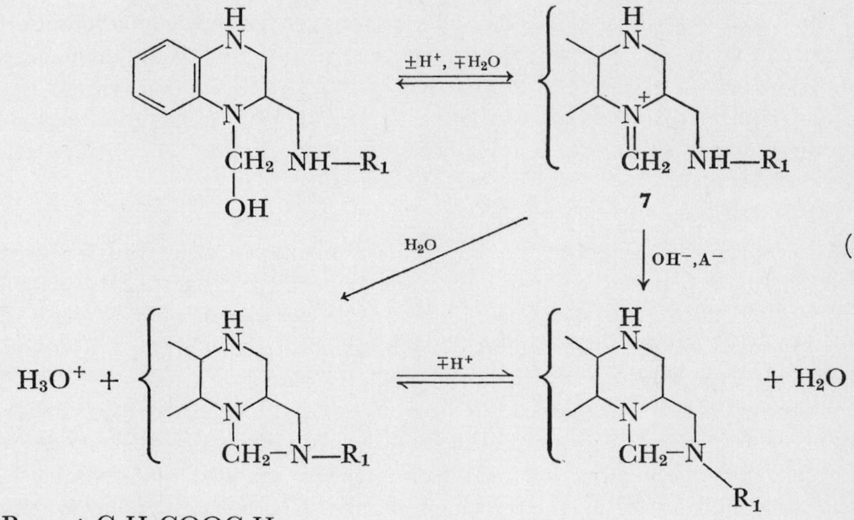

$$R_1 = p\text{-}C_6H_4COOC_2H_5 \tag{5}$$

The observation and identification of general base catalysis in imidazolidine formation suggests a general acid catalyzed mode of ring-opening, an attractive pathway for a biochemical process. Ring-opening can be viewed as a balance between two discrete but thermodynamically opposing steps: (1) protonation at the least basic nitrogen via a thermodynamically unfavorable equilibrium—it is possible that under certain conditions this step may be rate-controlling; and (2) expulsion of the more acidic protonated amine in order to yield the thermodynamically more stable iminium cation at the more basic amino site (N-1 versus N-10). This description should not be construed to mean that the actual mechanism is a stepwise process. The Brønsted β for the p-carbethoxytetrahydroquinoxaline model is suggestive of a concerted mechanism 8.

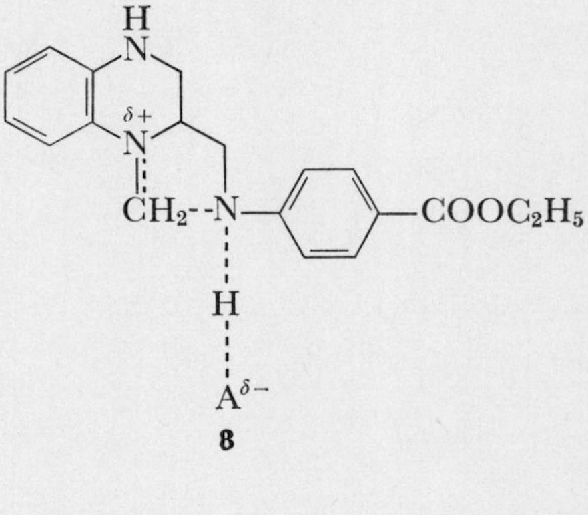

The inability to detect kinetically the iminium cation in the condensation of formaldehyde with the natural cofactor is probably due to the interfering ionization of the 4-hydroxy group that is manifested in the critical pH region required for detection. However, the successful synthesis of 5-methyl-H_4-folate from 5,10-methylene-H_4-folate in the presence of borohydride suggests the presence of the iminium species [28].

It is important to stress that for both of the above condensation reactions, one is observing the cyclization from the vantage point of a carbinolamine formed in a preequilibrium process, and hence the concentration of those species is under thermodynamic control. In other words, the ring-closure step involving the iminium cation is only kinetically detectable when steps involving amine-formaldehyde condensation followed by dehydration of carbinolamine are rapid and reversible relative to the cyclization step. Under these circumstances the concentration of the more stable carbinolamine (N-5 in the cofactor; N-1 in the tetrahydroquinoxaline model [29]) would dominate so that the observed kinetics might mainly reflect the ring closure reaction of the N-5 or N-1 iminium species. The argument is illustrated in Figure 2 in terms of a hypothetical free energy-reaction coordinate diagram. In essence the rate of cyclization of the N-5 iminium cation through T.S.-3 may be faster than dehydration of carbinolamine at N-10 through T.S.-1. The diagram illustrates the possibility—not revealed by the aforementioned kinetic studies—that ring opening of the cofactor may favor formation of the N-10-iminium cation species. If one views only the partitioning by ring opening of 5,10-methylene-H_4-folate to the respective iminium cations, the postulated behavior is typical of a reaction under kinetic versus

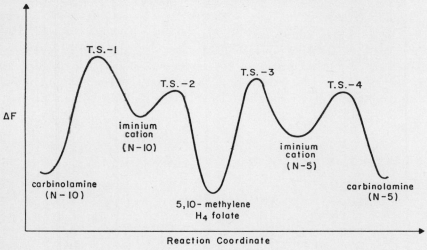

Figure 2

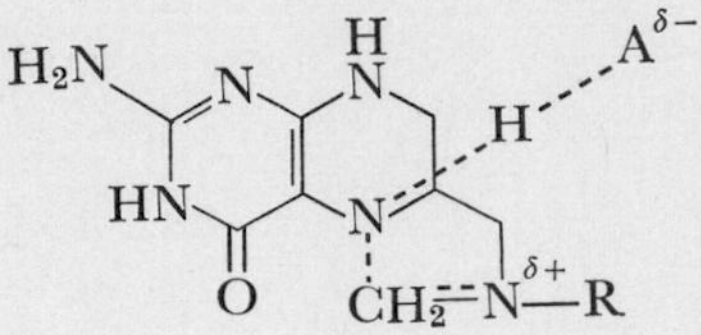

9

thermodynamic control. Such control has been demonstrated on the formyl level of oxidation for both reactions involving the cofactor and the analog. Indeed, the positioning of the catalytic groups on the enzyme itself may preclude opening to the N-5 cation. A likely process for the first step of enzymic reactions involving 5,10-methylene-H_4-folate is given in **9**. The concept of general acid catalysis might not be invalidated, although changes in the effectiveness of catalysis (decreasing β) are anticipated. One also may surmise that the magnitude of ΔpK_a (N-5 versus N-10) for H_4-folate approximates the optimal value needed to delicately balance ring integrity— in order to insulate the biological system from free formaldehyde—against the free energy changes required for ring opening in the N_{10} direction.

The above data collectively argue for the probable existence of an iminium cation in the transfer and ring-opening reactions of the natural cofactor. Indeed, such a species had been proposed earlier by both Jaenicke and Huennekens as participating in the hydroxymethyl transfer reactions of the H_4-folate cofactors [30, 31]. Moreover, iminium cations long have been implicated as intermediate species in a variety of condensation reactions proceeding in acid media [32] (Mannich, Leuchart, and Polonovski) and more recently have been directly observed in strong acid solution with nmr techniques [33, 34].

Against this background let us consider now the possible mechanisms for the enzyme thymidylate synthetase that catalyzes the transformation shown in (6). The pertinent findings to be considered may be summarized as follows. (1) Kinetic studies focusing on analog and product inhibition patterns of the Ehrlich ascites synthetase are consistent with an ordered

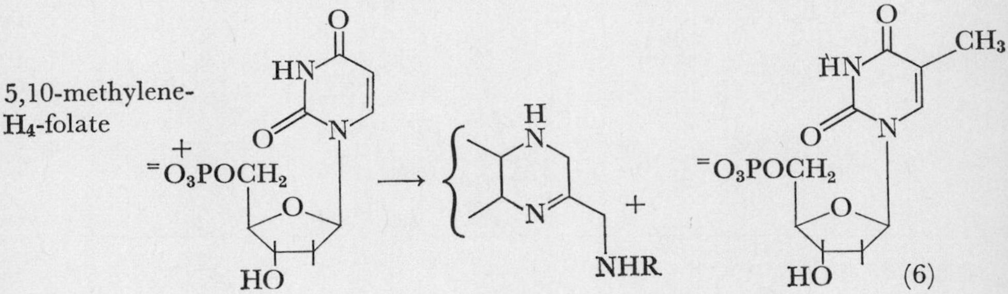

5,10-methylene-H_4-folate

(6)

sequential mechanism in which the 5,10-methylene-H_4-folate combines first with the enzyme [35]. (2) Similar investigations with the synthetase isolated from chicken embryo also are in accord with an ordered sequential mechanism, although in this case deoxyuridine monophosphate apparently binds first [36]. (3) Thymidylate synthetase from *E. coli* catalyzes the exchange of hydrogen at *C*-5 of the nucleotide in the presence of H_4-folate [37]. (4) Although 5-hydroxymethyl-2'-deoxyuridine monophosphate is a competitive inhibitor with respect to 2'-deoxyuridine monophosphate, it has no substrate properties [38, 39]. (5) The catalytic activity of various highly purified preparations of thymidylate synthetase is dependent on the presence of an essential sulfhydryl group [40–42]. (6) The hydrogen at *C*-6 of the cofactor is incorporated into the methyl group of the thymidylate product [36, 43, 44]. Despite this accumulation of information one still is allowed considerable latitude in formulating a reaction mechanism. One possibility consistent with both the model and enzymic studies is illustrated in (7) and (8), where EB, EBH, and ESH (ES⁻) represent enzymic base, acid, and thiol groups, respectively.

Let us now amplify our description of certain of the above steps. The postulated thiol-catalyzed exchange of hydrogen at *C*-5 of the pyrimidine ring has considerable precedent in the observed exchange of *H*-5 of uridine in basic media [45], the analogous hydroxide ion catalyzed deuterium exchange reaction of 1-methyluracil-5-*d* [46], and the glutathione catalyzed isotope exchange of uridine [47]. All these reactions are thought to proceed through rate-determining 1,4 addition of the nucleophile to the 5,6 double bond of the *N*-1 substituted pyrimidine, followed by protonation-deprotonation and elimination of the nucleophilic species (9). Variations of the

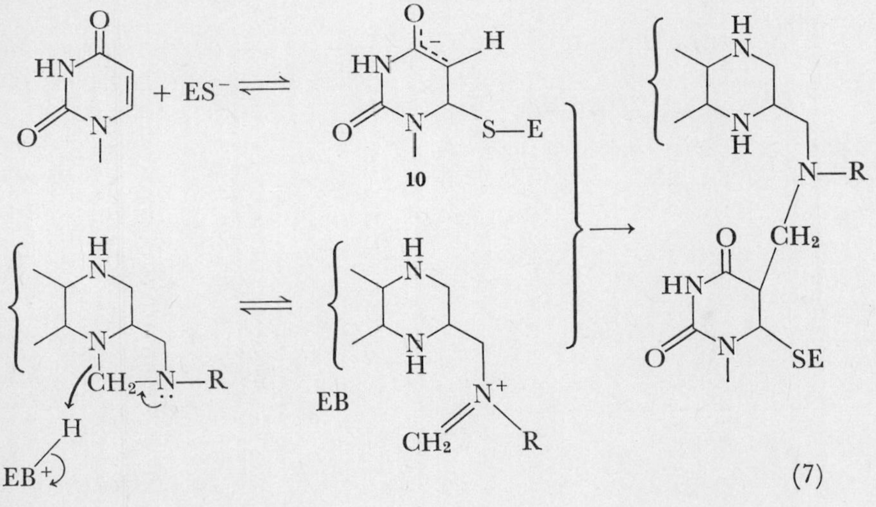

(7)

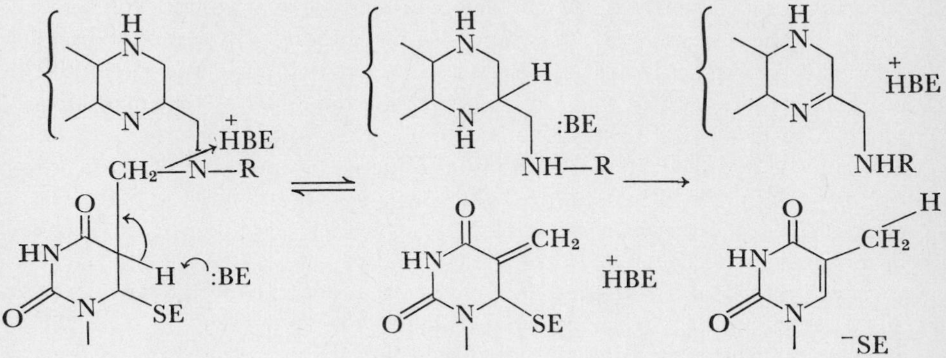

(8)

latter mechanism have been offered [47]. One such mechanism involves 1,4 addition of the conjugate acid (HB⁺), which is, at present, inconsistent with the kinetic data where the rate equation is first-order in both nucleophile (B) and un-ionized uridine. If, in the enzymic reaction, the nucleophile approaches from only one side of the pyrimidine ring, the observation of exchange at C-5 requires that C-5 be accessible to solvent from both sides. Since the rate of exchange in the complete enzymic system is only 5–10% of that of the overall reaction, it seems desirable to determine if isotopic exchange at C-5 is enzyme catalyzed. Presumably, an inhibited enzyme would also catalyze exchange if catalysis is remote from the active site. Finally, it should be stressed that participation of N-1 of the pyrimidine ring in exchange and condensation reactions via species such as **11a** is probably only encountered when N-1 is capable of ionization [48].

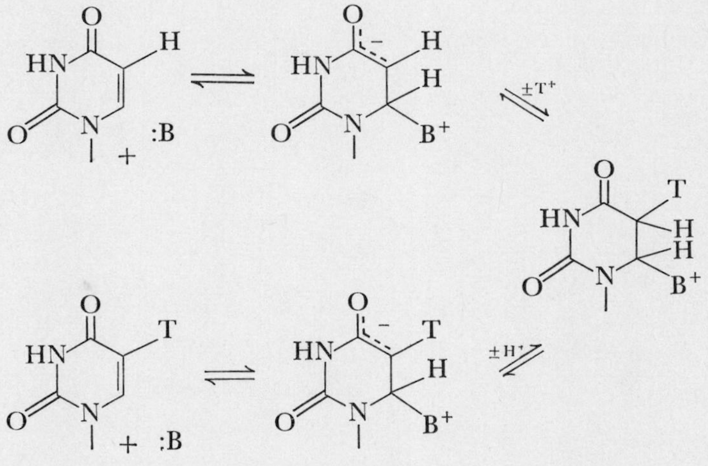

(9)

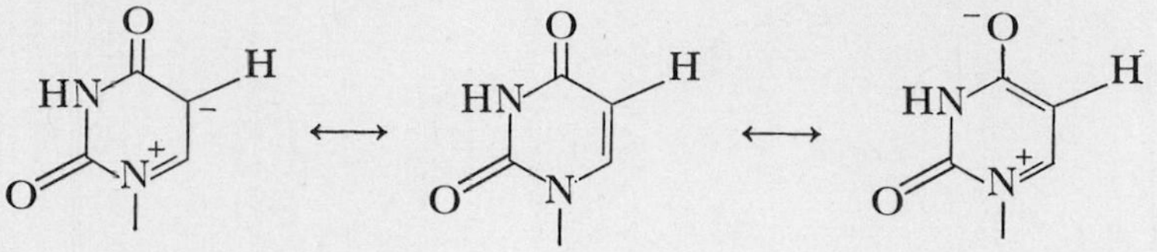

11a   11b   11c

An intermediate similar to the enzyme-bound cofactor-pyrimidine adduct has previously been described by Friedkin and Kornberg [49], although in that case no covalent linkage to the enzyme was postulated, and alkylation was at N-5. The proposed requirement for nucleophilic catalysis in the condensation may explain the failure to detect the accumulation in solution of such a species, although this is by no means the only rationale.

Decomposition of the proposed adduct is depicted simply as general catalysis of an elimination reaction of which there are numerous examples. It is interesting to consider the behavior of 5-thyminyl-H_4-folate (**12**) and the corresponding 10-isomer (**13**) at neutral pH. Whereas the 10-isomer spontaneously decomposes to hydroxymethyluracil and the parent cofactor, the 5-isomer is recovered unchanged [50]. This observation is in accord with the relative instability of other N-10 substituted H_4-folates as compared with their N-5 isomers and may be a very preliminary indication of the kinetic advantage for employing the N-10 isomer. The mechanism for decomposition of **13** probably involves initial addition of water to C-5 and follows a mechanism similar to (10). In this sense the breakdown of **13** is not strictly analogous to that proposed in (8) unless the quinone methine species has —OH at C-5 rather than —SE. Loss of ES⁻ prior to elimination is an alternative pathway.

Finally, let us consider the exocyclic methylene intermediate formed in the elimination reaction. Similar species have been implicated in the

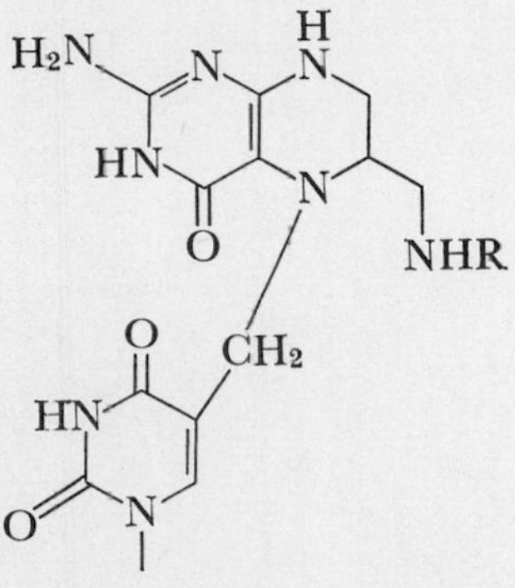

12 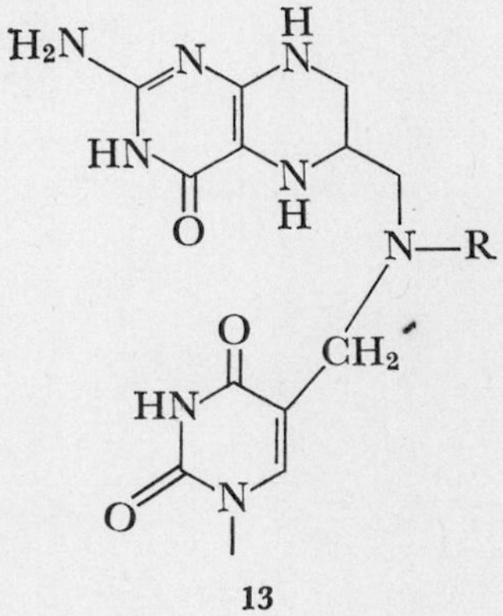   13

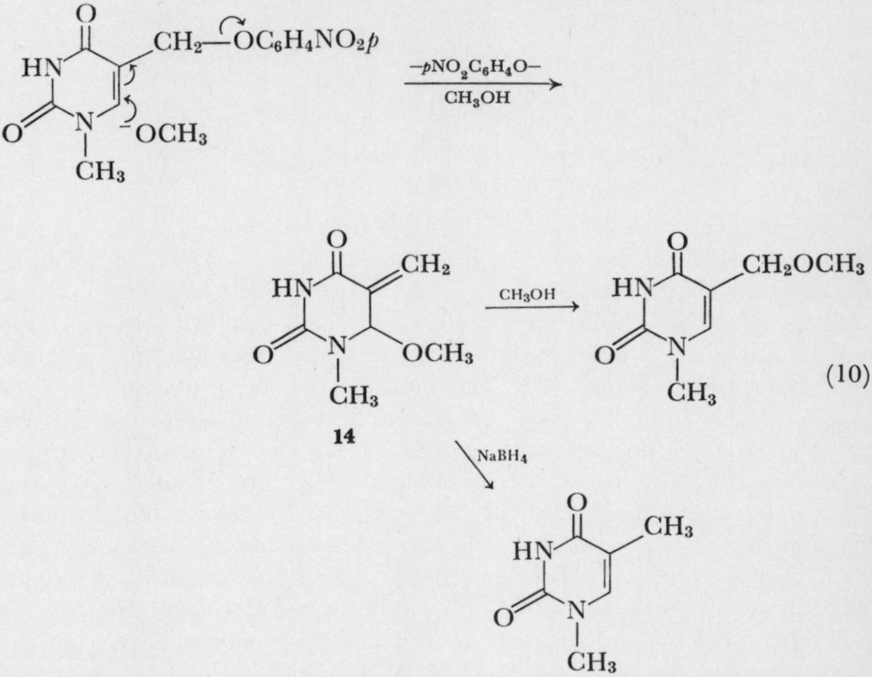

(10)

hydrolysis of 5-trifluoromethyluracil and its *N*-alkylated derivatives and in nucelophilic substitution reactions of 5-acetoxymethyl- and 5-(*p*-nitrophenoxy)-methyluracils and their *N*-alkylated isomers [49, 51]. When ionization is precluded, as in the 1-methyl derivatives, the reactions proceed through addition of the nucleophile to the 6-position of the heterocycle to generate the 5-methyleneuracil intermediate, **14**. In the presence of sodium borohydride the course of the reaction is diverted to give 1-methylthymine as the sole product. The irreversible inhibition of thymidylate synthetase by 5-trifluoromethyluracil has been attributed to capture of a similar reactive intermediate by a proximate enzyme nucleophilic group. Thus it appears that the exocyclic methylene uracil species may indeed be an intermediate possessing suitable characteristics for the final oxidation-reduction step.

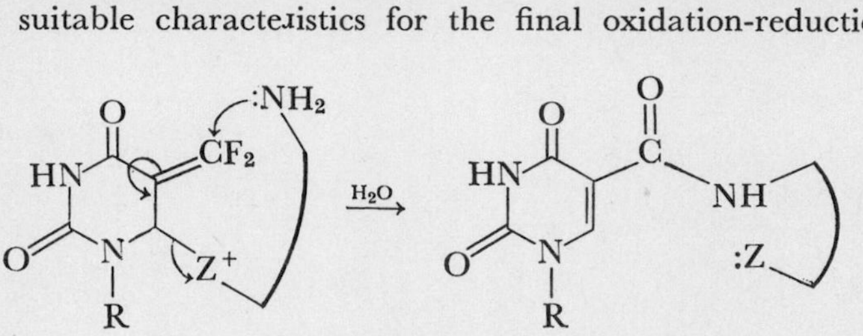

(11)

R = deoxyribose-5'-phosphate

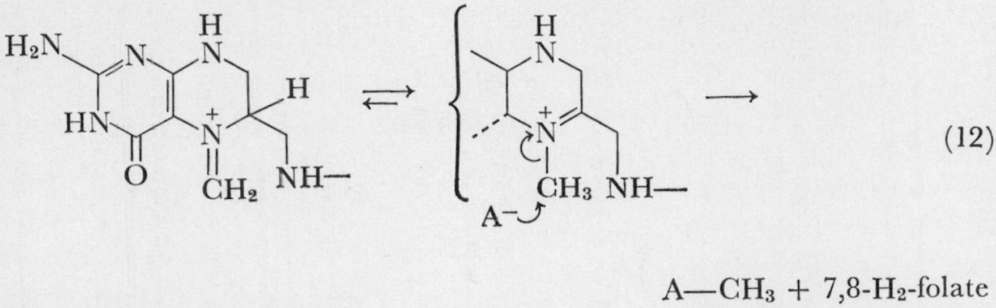

$$A—CH_3 + 7,8-H_2\text{-folate} \tag{12}$$

The timing of the molecular events in the latter process is obscure; one may envision electron transfer followed by proton transfer occurring through the intermediacy of a charge-transfer complex, hydride transfer, or, alternatively, addition-elimination processes involving N-5. In any event the hydrogen at C-6 must be retained within the enzyme complex. The reaction in an analogous carbon system, that is, 1,4 addition to butadiene of hydrogen derived from 1,2 dehydrogenation of ethane, is symmetry forbidden, but this argues only against a fully concerted process.

Variations on the above mechanism may be written, of course; for example concerted instead of sequential processes. However, two possibilities are worthy of comment since there is evidence presently available that suggests their operation to be less likely. One such mechanism involves tautomerization of the iminium cation to 5-methyl-7,8-dihydrofolate that would then function as a methylating species [52]. The 1,3 suprafacial rearrangement is a symmetry-forbidden process, so that presumably the reaction—if concerted—would proceed in an antarafacial manner, necessitating the involvement of more than one basic group on the enzyme [53]. More compelling evidence against (12), however, is the low basicity of C-6, as manifest in the negligible exchange rates of such hydrogens in relevant model iminium cations [54]. Based on this line of reasoning, it is also doubtful that the cofactor first methylates the enzyme, the methyl group in turn being transferred to the uridyl moiety [55].

The hydroxymethyl transfer reactions, which include the serine-glycine interconversion and the formation of hydroxymethyldeoxycytidylate, may be profitably considered together.

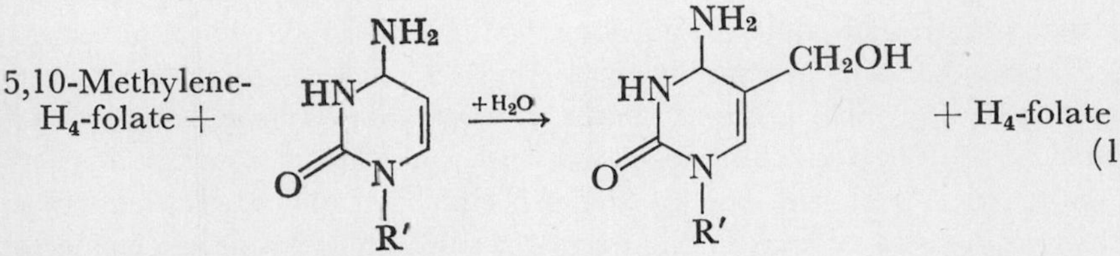

5,10-Methylene-
H_4-folate +

$$\tag{13}$$

where $R' =$ deoxyribose-5'-phosphate

$$5,10\text{-Methylene-}H_4\text{-folate} + \underset{\overset{|}{NH_2}}{CH_2}\text{---COOH} \underset{\rightleftharpoons}{\overset{\pm H_2O}{}}$$

$$\underset{\overset{/}{NH_2}}{\overset{CH_2OH}{\underset{\backslash}{CH}}}\text{---COOH} + H_4\text{-folate} \qquad (14)$$

Comparatively little is known about the mechanism of the former reaction except that the enzyme catalyzes a H_4-folate dependent exchange between the hydrogen at C-5 of deoxycytidylate and water [56]. The exchange reaction proceeds more slowly in the presence of 5,10-methylene-H_4-folate. One may again ascribe the exchange reaction to an addition reaction across the 5,6 double bond of the pyrimidine ring with the cofactor required to produce a catalytically active ternary enzyme-cofactor-pyrimidine complex. The decrease in exchange with the requisite 5,10-methylene-H_4-folate may reflect partitioning of the activated pyrimidine species between exchange and conversion to product.

The highly purified rabbit transhydroxymethylase has been extensively studied by Jenkins, Schirch, and their co-workers [57]. The enzyme, which requires pyridoxal phosphate for full enzymic activity, catalyzes several reactions other than (14), including:

$$\alpha\text{-Methylserine} + H_4\text{-folate} \rightleftharpoons \text{D-alanine} + 5,10\text{-methylene-}H_4\text{-folate} \qquad (15)$$

$$\text{L-threonine} \rightleftharpoons \text{glycine} + \text{acetaldehyde} \qquad (16)$$

$$\text{Allothreonine} \rightleftharpoons \text{glycine} + \text{acetaldehyde} \qquad (17)$$

$$\text{D-alanine} + \text{holoenzyme} \rightleftharpoons \text{pyruvic acid}$$
$$+ \text{pyridoxamine phosphate} + \text{apoenzyme} \qquad (18)$$

Note that H_4-folate is required for only the serine-glycine and α-methyl-serine-D-alanine interconversions. In an important investigation with this enzyme, Jordan and Akhtar, and Besmer and Arigoni extended an earlier observation that serine transhydroxymethylase catalyzes exchange of the α-hydrogen of glycine, the rate of which was increased in the presence of H_4-folate [58, 59, 59a] but decreased with increasing concentrations of formaldehyde. With the use of stereospecifically labeled glycine, these authors demonstrated that in the overall reaction that converts glycine to L-serine, the cleavage of the C—H bond of glycine and the formation of a

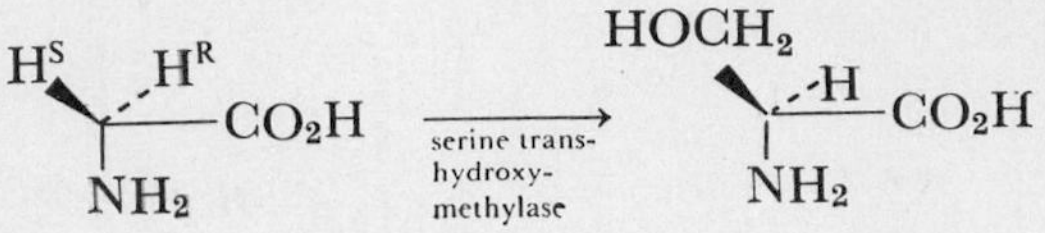

$$\underset{NH_2}{\overset{H^S\quad H^R}{\diagdown}}\!\!\!-CO_2H \quad \xrightarrow[\substack{hydroxy-\\methylase}]{serine\ trans-} \quad \underset{NH_2}{\overset{HOCH_2}{\diagdown}}\!\!\!\overset{-H}{\diagdown}-CO_2H \qquad (19)$$

new C—C bond of serine occur with retention of configuration and loss of only the S-hydrogen (19). The identical S-hydrogen is lost during threonine biosynthesis. Furthermore, apparently no hydrogen is lost from the methylene bridge of the cofactor.

It is not our intention to discuss the chemistry of pyridoxal phosphate [60]. May it suffice to state that pyridoxal phosphate probably promotes the exchange reaction of glycine through Schiff-base formation and that H_4-folate by binding enhances the deprotonation of this initial complex [58]. For both the hydroxymethyldeoxycytidylate and serine transhydroxymethylase, we now have the key intermediates possessing carbanionic character for the requisite condensation process, that is, **10** as depicted for thymidylate synthetase and **15b**.

The question is whether the iminium cation, employed earlier, is also relevant here. Following the condensation of **15b** with an iminium cation either at N-5 or N-10, let us survey the subsequent possible pathways, (21) and (22).

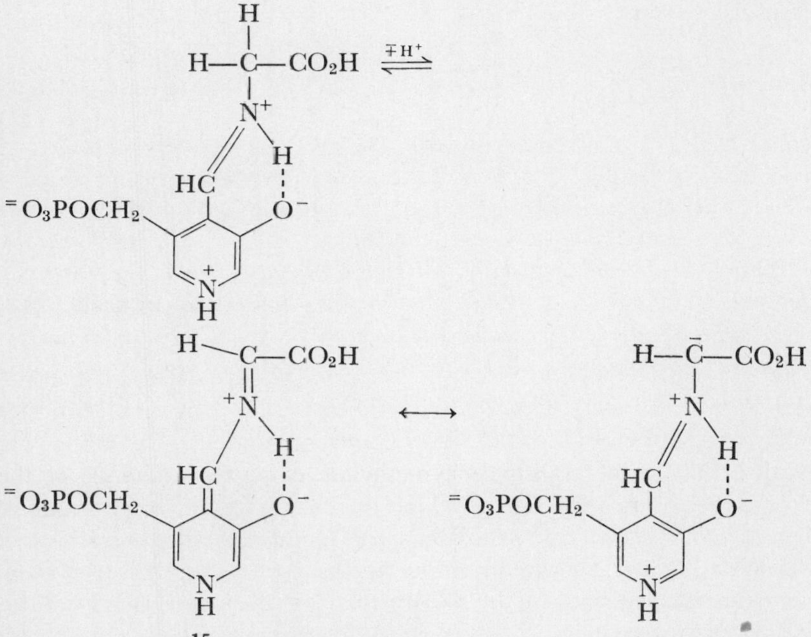

$$(20)$$

15a   15b

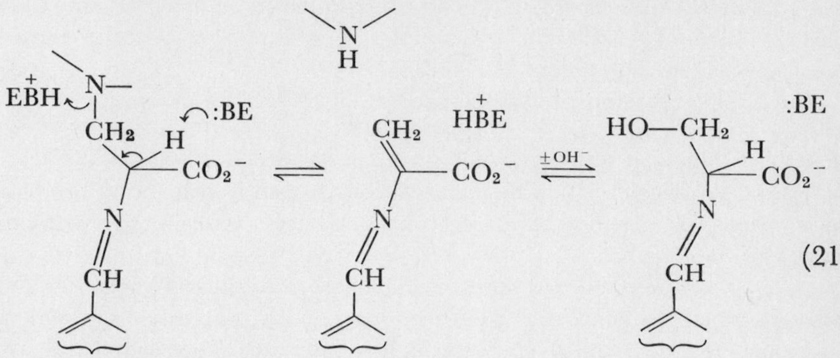

$$
\text{(21)}
$$

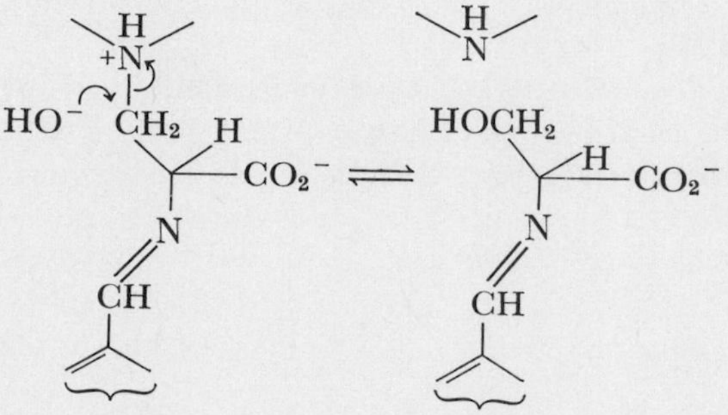

$$
\text{(22)}
$$

Mechanism (21) is inconsistent with the loss of only one hydrogen to solvent in the course of the reaction unless the enzyme catalyzes the hydrolysis of the acrylic acid intermediate without proton exchange with the media. However, the conversion of α-methylserine to alanine obviously cannot follow this route. Mechanism (22), although in accord with the data, not only features a difficult S_N2 displacement but is contrary to the reactions of carbonyl compounds, which generally occur by an addition-elimination pathway. A third mechanism (23) views 5,10-methylene-H_4-folate functioning simply as a carrier of formaldehyde in these hydroxymethyltransferase reactions. This mechanism likewise is not without its difficulties since it must

$$
\text{5,10-Methylene-}H_4\text{-folate} \underset{}{\overset{\pm H_2O}{\rightleftharpoons}} H_4\text{-folate} + CH_2O \underset{}{\overset{\pm AH}{\rightleftharpoons}} \text{A-}CH_2OH \quad \text{(23)}
$$

be argued that the formaldehyde is not free to rotate before condensation in order to rationalize the evidence that the transfer of the one carbon unit from tritium labeled 5,10-methylene-H_4-folate to serine is stereospecific. Thus, a significant experiment would be to determine whether the presence of H_4-folate effects the ratio of threonine to allothreonine since one anticipates that acetaldehyde also might be restricted in its rotation within the enzyme-substrate-cofactor complex.

The latter (23) is an attractive speculation, especially in view of the lack of requirement for H_4-folate in other reactions catalyzed by serine hydroxymethyl transferase. Furthermore, one can imagine that the absence of a requirement for an oxidation-reduction step as found with thymidylate synthetase obviates the need for generating an unsaturated reducible methylene species. Such an intermediate is more readily produced through condensations involving iminium cations than through dehydrations of the corresponding alcohol derived from formaldehyde condensation. This hypothesis is also consistent with the observation that 5-hydroxymethyl-2'-deoxyuridylate, although a competitive inhibitor with respect to 2'-deoxyuridylate, has no substrate properties in the thymidylate synthetase reaction. The aforementioned mechanism may prove to be general for hydroxymethyl-transferases and distinct from that for thymidylate synthetase.

4 TRANSFER OF THE ONE-CARBON UNIT AT THE OXIDATION LEVEL OF FORMATE

Most of the enzymic reactions involving H_4-folate in the metabolism of formate-derived carbon units can be discussed under one of three headings defined by reaction type and/or probable mechanism. The classifications chosen are the following: (1) reactions involving amidine-orthoamide equilibria, (2) formyl transfer, and (3) activation of formate and formyl-H_4-folate.

4.1 Amidine-Orthoamide Equilibria

At least four of the folate-dependent enzymic reactions thus far reported involve amidines as substrates and/or products. The stoichiometry of the reactions is indicated in equations (24)–(27) along with the systematic names of the enzymes. In each case, one might envision aminolysis of an amidine as the initial step of the transformation. Because this step must lead to a triaminomethane or orthoamide intermediate, consideration of the formamidine-orthoformamide equilibrium might afford comment on the mode of enzymic catalysis.

Although orthoamides in general have been considered to be both thermally and hydrolytically labile with respect to decomposition to their amidine and amine constituents [61], recent investigations have demonstrated the existence of orthoformamides as steady-state intermediates in the aminolysis of amidines in aqueous solutions [62]. Central to arguments concerning enzymic catalysis involving orthoformamides would be a detailed description of their hydrolytic behavior, especially with respect to

5-Formimino-H_4-folate + H^+ $\rightleftharpoons$ 5,10-methenyl-H_4-folate + NH_3
5-Formiminotetrahydrofolate ammonia-lyase (cyclizing)
EC 4.3.1.4.

$$(24)$$

Formiminoglutamic acid + H_4-folate $\rightleftharpoons$ 5-formimino-H_4-folate
+ glutamic acid
N-formimino-L-glutamate:tetrahydrofolate 5-formimino-
transferase EC 2.1.2.5.

$$(25)$$

N-formiminoglycine + H_4-folate $\rightleftharpoons$ 5-formimino-H_4-folate +
glycine
N-formiminoglycine:tetrahydrofolate 5-formimino-transferase
EC 2.1.2.4.

$$(26)$$

5,10-Methenyl-H_4-folate + H_2N-CH_2-$CONH$-ribose-P $\overset{\pm H_2O}{\rightleftharpoons}$ H_4-
folate + OHC-HN-CH_2CONH-ribose-P.
5,10-Methenyl-H_4-folate: 2-amino-N-ribosylacetamide-5′-phosphate
transformylase.

$$(27)$$

relative rates of amine expulsion as a function of pK_a. As proposed earlier for acid catalyzed ring opening of 5,10-methylene-H_4-folate, the favored route for decomposition of orthoamides is presumably dictated by the relative importance of two thermodynamically opposing situations, in that protonation and expulsion of the most basic nitrogen lead to the least stable amidine, and conversely, formation of either of the more stable amidines requires protonation at a less thermodynamically favorable site. The same considerations can be applied to the decomposition of a tetrahedral addition compound in amidine hydrolysis. In the latter case, the stabilities of the incipient amides are not reflected in the respective transition states, and the more basic nitrogen is expelled [63].

A model reaction designed to comment on the hydrolytic behavior of orthoformamides and the related formamidines is represented schematically in (28). Hydrolysis of the 5-formimino analog **16** follows first-order kinetics and exhibits a pH-dependent product distribution ranging from ~80 % **21**, 20 % **19** at pH 7 to the inverse distribution at pH 9, owing to increasing hydroxide ion catalyzed hydrolysis of the acyclic formamidine **16b** at higher pH. Although (28) demonstrates the facility of formamidine interconversion via the orthoformamide **17** as well as the orientation of amidine hydrolysis, the data are preliminary and at the time of writing may be satisfactorily reconciled without inclusion of the N-10 formamidine **18a–b** as an intermediate. As indicated, the kinetics do not distinguish whether formation or breakdown of **17** is rate limiting. In light of the near structural degeneracy of the equilibria interrelating **17** and the two acyclic formamidines **16** and **18**, further work is in progress to determine whether or not **18** is formed and in turn establish the preferred routes for decomposition of the orthoformamide. It should be noted that acid catalyzed decomposition to products of the proposed orthoformamide intermediates relevant to (24)–

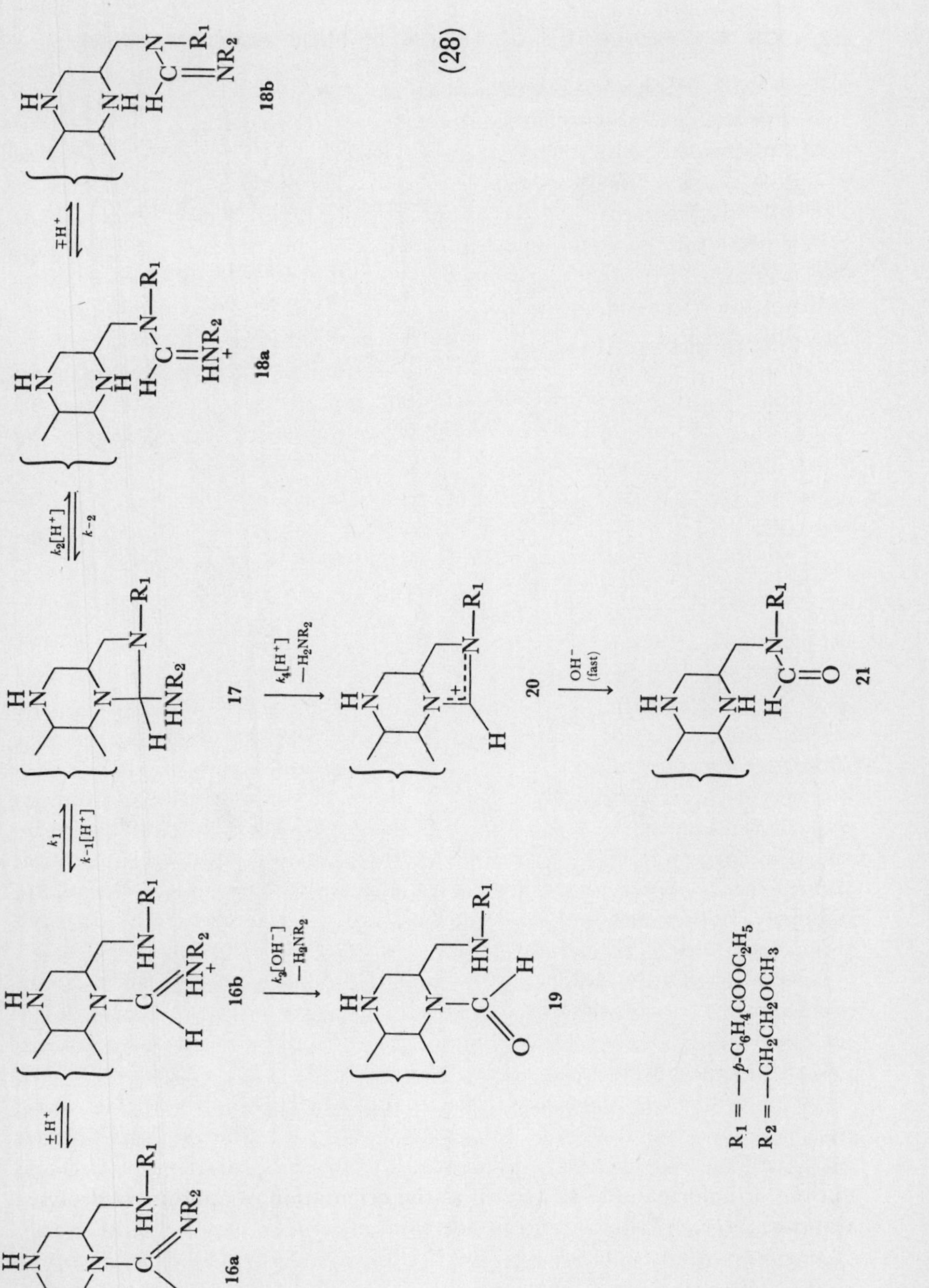
(28)
16a
16b
17
18a
18b
19
20
21
±H⁺
k₁
k₋₁[H⁺]
k₂[H⁺]
k₋₂
∓H⁺
k₃[OH⁻]
−H₂NR₂
k₄[H⁺]
−H₂NR₂
OH⁻
(fast)
R₁ = —p-C₆H₄COOC₂H₅
R₂ = —CH₂CH₂OCH₃

(27) would involve monoprotonation at the most thermodynamically favorable site (most basic nitrogen)[f] for all but that of (27).

Mechanistically, the cyclodeaminase reaction (24) is possibly the most straightforward of the four. Conversion of 5-formimino-H_4-folate to 5,10-methenyl-H_4-folate can be effected nonenzymically at low pH and has been used to monitor the formimino transferase activities of (25) and (26) since the latter exhibits a λ_{max} at 353 nm (ε 25,000) at pH 2 [64]. If the mechanism of the enzymic reaction resembles that of the nonenzymic pathway, one can argue from the results of (28) that the key steps are simply formation of the orthoformamide **22** and subsequent loss of ammonia [65]. As intimated in later discussion [cf. (48)] it is possible that, in both the model and enzymic reactions, deprotonation of the N-10 nitrogen might provide a major energy barrier in the net cyclization.

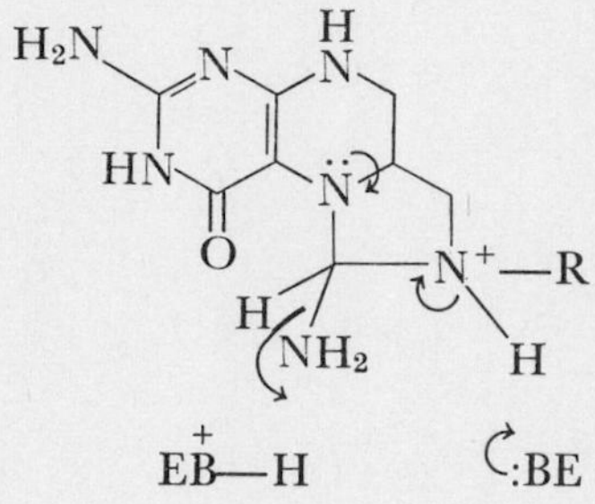

22

Consideration of the formimino transfer reactions of (25) and (26) and related model studies implicates an additional role of the protein in catalysis of these reactions. Formation of the orthoformamide **17** is apparently greatly facilitated by intramolecular juxtaposition of the attacking amine, because aminolysis of **20** does not effectively compete with hydrolysis in aqueous solution [66]. In the absence of enzymic catalysis, hydrolysis of N-formiminoglutamate or N-formiminoglycine—presumably through their respective protonated acyclic formamidinium salts [63]—would proceed at a rate greatly in excess of that for aminolysis by the N-5 position of H_4-folate. In order to retain the orthoformamide hypothesis one must invoke a local concentration effect conferred by enzymic juxtaposition of the N-formimino and H_4-folate substrates. The previous suggestion of general catalysis in the formation and decomposition of the orthoamide **22** presumably would apply to the acyclic species **23** as well.

An alternative mode of catalysis for the net transfer of the formimino moiety should be considered. This mechanism obviates the necessity for direct aminolysis by generating an enzyme-bound formimino intermediate. Some precedent for this possibility is the finding that an isothiourea-enzyme

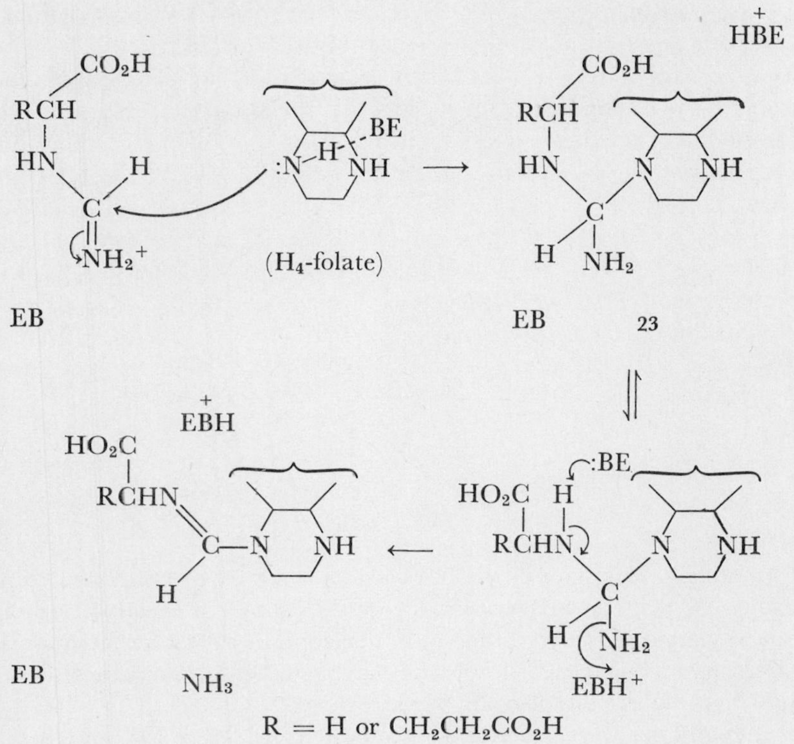

$$
\begin{array}{ccc}
& & \overset{+}{HBE} \\
\underset{EB}{\overset{CO_2H}{\underset{\underset{\overset{\displaystyle C}{\underset{NH_2^+}{\|}}}{HN\quad H}}{RCH}}} \qquad
\overset{H}{\underset{(H_4\text{-folate})}{:N\diagdown \diagup NH}}{\overset{\diagup BE}{}} & \longrightarrow &
\underset{EB\quad 23}{\overset{CO_2H}{\underset{\overset{\displaystyle C}{\underset{NH_2}{H}}}{\underset{HN\quad N\quad NH}{RCH}}}}
\end{array}
$$

(29)

$$
\begin{array}{ccc}
\overset{+}{EBH} & & \overset{:BE}{} \\
\underset{EB\quad NH_3}{\overset{HO_2C}{\underset{H}{\underset{RCHN\diagdown}{}}}\;C-N\quad NH} & \longleftarrow &
\underset{EBH^+}{\overset{HO_2C\ H}{\underset{\overset{\displaystyle C}{\underset{NH_2}{H}}}{RCHN\diagdown\ N\quad NH}}}
\end{array}
$$

$$R = H \text{ or } CH_2CH_2CO_2H$$

intermediate is involved in a transamidinase reaction [67], suggesting that a thioimidate-enzyme intermediate might operate in formimino transfer reactions. The two steps then would be thiolysis of the formamidine, followed by aminolysis of the resulting enzyme-bound thioimidate (30). Unfortunately, there is little in the way of model reactions to draw upon in this case, and it is not obvious if this route would afford a kinetic advantage.

The so-called transformylase reaction of (27) has received surprisingly little attention in light of the fact that it lies directly on the pathway for purine biosynthesis. The enzyme from acetone-dried chicken liver was purified 58-fold by Buchanan and co-workers and was shown to be specific for both 5,10-methenyl-H_4-folate and glycinamide ribonucleotide-5'-phosphate [68]. Despite the lack of data on the enzymic reaction, discussion of possible mechanisms is perhaps warranted because a similar nonenzymic transfer has been demonstrated and partially characterized with respect to mechanism [69].

In keeping with the mechanisms discussed above one might expect condensation of the α-amino group of glycine with the cationic 5,10-methenyl-H_4-folate to be the first step in the transfer reaction. Nonenzymic hydrolysis

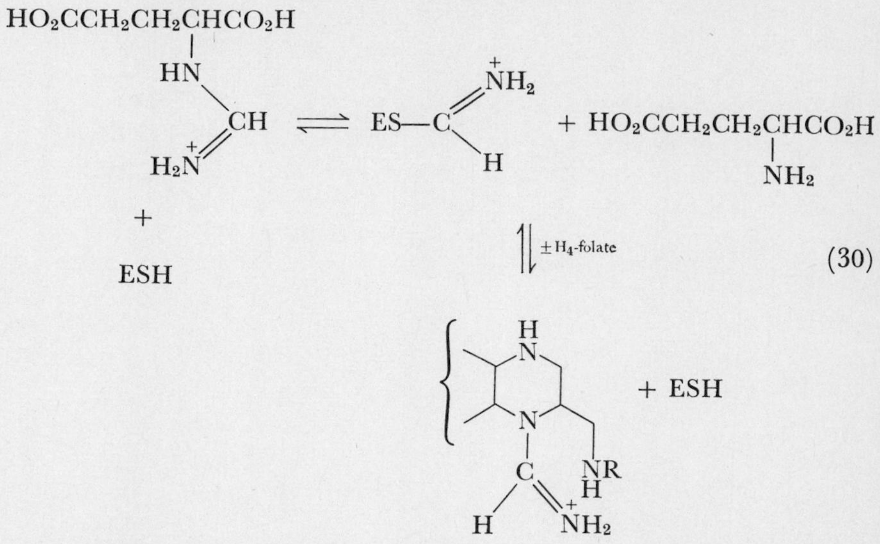

$$HO_2CCH_2CH_2CHCO_2H$$

$$\text{(30)}$$

of 5,10-methenyl-H_4-folate in the presence of a variety of buffers including hydrazine (20–80% free-base form) proceeds without detectable accumulation of any intermediate resulting from nucleophilic attack on the substrate [70]. Here again, one must assign to the enzyme the task of forming a ternary complex favoring the bimolecular condensation.

The hypothetical orthoformamide intermediate **24** must now be reorganized and/or hydrolyzed with scission of both C—N bonds involving H_4-folate. Reference again to (28) reveals that uncatalyzed decomposition of this intermediate would lead either to starting material or to the substituted N-5-formimino-H_4-folate **25** or the N-10 isomer. Subsequent hydrolysis of either acyclic amidine, however, would probably expell the

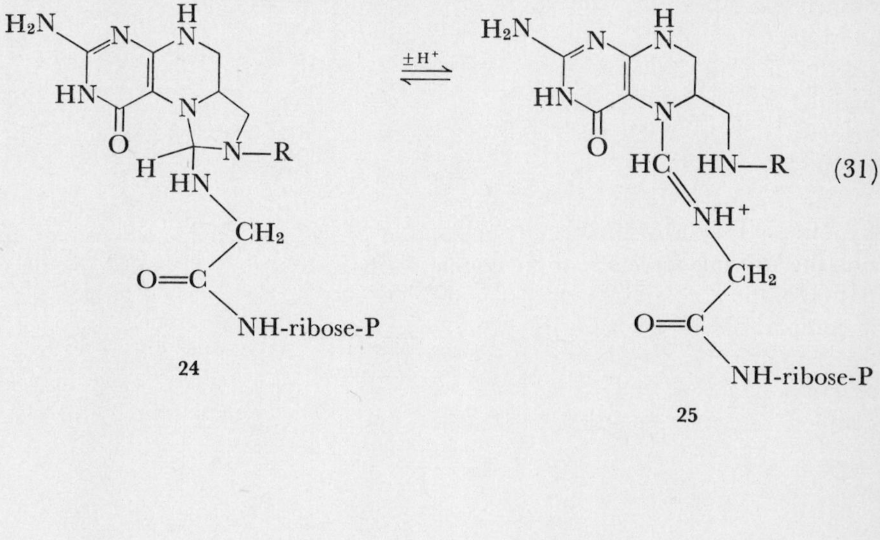

$$\text{(31)}$$

24

25

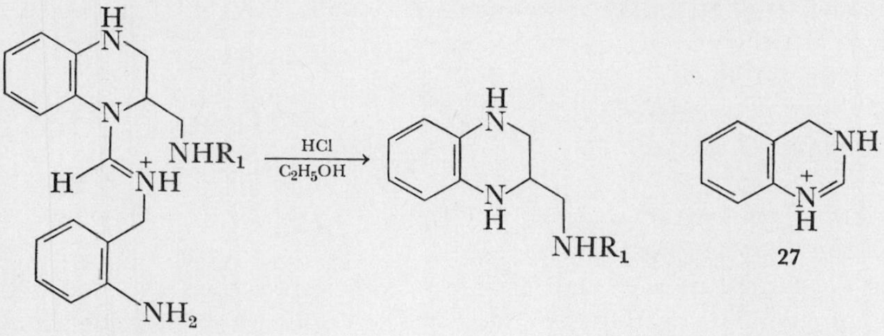

(32)

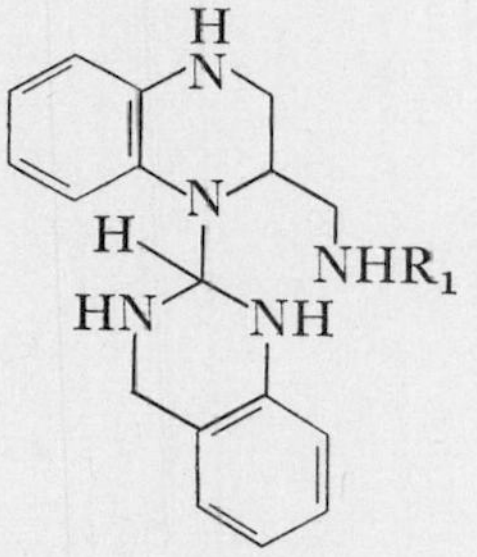

glycinamide moiety preferentially, yielding no net transfer products. Whereas it would suffice to invoke enzymic mediation of the prototropy leading to transfer of the formate carbon—although in a thermodynamically unfavorable direction—extension of the model studies of (28) provides evidence for an alternative pathway.

Although condensation of the formamidinium salt 29 with o-aminobenzyl-amine in ethanol should lead to the formamidine 26, attempts to isolate the adduct have been unsuccessful due to subsequent rapid conversion of the presumed material to the model diamine and 3,4-dihydroquinazoline 27 [69]. The observed change in reactivity and orientation can best be explained by invoking nucleophilic participation of the aniline nitrogen and formation of the orthoformamide 28. Decomposition of 28 with expulsion of the protonated N-1 nitrogen would lead to the observed products.

The facility exhibited by the nonenzymic reaction gives rise to the hypothesis that a neighboring nucleophile provides a kinetic advantage if not the driving force for the enzymic formate transfer. Although participation of the amide function of the glycinamide ribonucleotide is implicated, a comparable mechanism might envision nucleophilic participation of a functional group on the enzyme. It should be noted that the enzymic reaction has been monitored either by release of free H_4-folate or by a coupled enzymic reaction involving N-formylglycinamide ribonucleotide phosphate, neither of which would implicate the unstable 4(5)-hydroxy-imidazole as the initially released product [68].

4.2 Formyl Transfer

Several enzymic reactions involving formylated tetrahydrofolic acid have been demonstrated and characterized to varying degrees. Intimately related to formyl transfer and formate metabolism in general is the small group of reactions effecting the interconversion of the three principal forms of formate-carrying H_4-folate. Accordingly, these reactions have received much attention, especially from the model standpoint. The transformations in question are shown in (33). Extensive use of models in the study of these reactions presumably resulted from the observations that the transformations can be effected nonenzymically and that the natural cofactors involved are unstable with respect to air oxidation. Specific discussion of the reactions requiring ATP is deferred to the section on activation of formate.

Shive et al. originally described the nonenzymic conversion of N-5- and N-10-formyl-H_4-folate to 5,10-methenyl-H_4-folate in the presence of acid as well as the base catalyzed hydrolysis of 5,10-methenyl-H_4-folate to give N-10-formyl-H_4-folate as the product of kinetic control and N-5-formyl-H_4-folate as the product of thermodynamic control [71]. With the aid of an

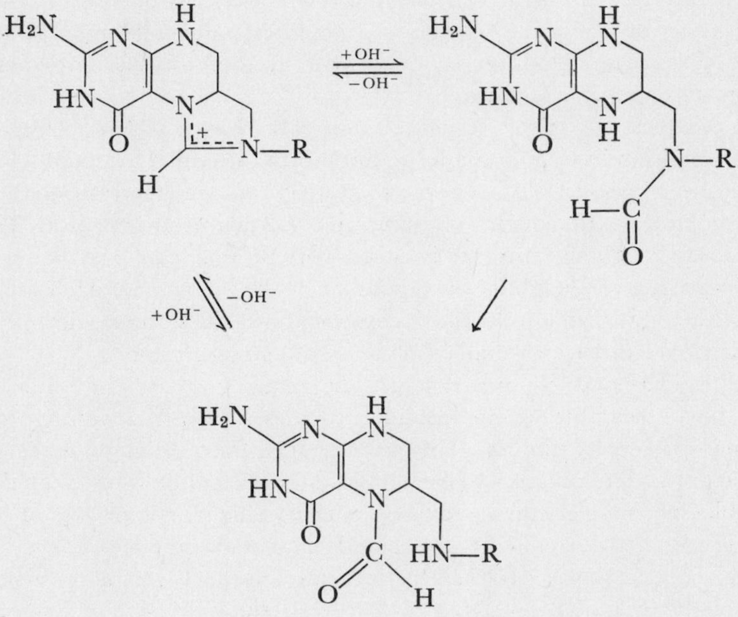

(33)

extensive model study, Robinson and Jencks later presented a detailed description of the mechanism for catalysis of N-10-formyl formation from 5,10-methenyl-H_4-folate [70, 72]. Thus at low buffer concentration (where breakdown of the tetrahedral addition intermediate is rate limiting) the reaction follows the rate law of (34) in which S represents the 5,10-methenyl-H_4-folate, k_0 is the spontaneous reaction rate constant, B is hydroxide ion

$$k_{\text{obs}} = v \frac{1}{[S]} = k_0 + k_1 [B] + k_2 [B][OH^-] \qquad (34)$$

or base form of the buffer, and k_1 and k_2 are rate constants for the terms first- and second-order in base respectively. A mechanism involving general base catalyzed breakdown of the cationic intermediate formed by the addition of the elements of water to 5,10-methenyl-H_4-folate was proposed with respect to the term first-order in base (35). The k_2 term is assigned to the mechanism of equation (36) in which breakdown of the anionic intermediate is facilitated by proton donation from the conjugate acid of the catalyst.

A model study related to the interconversions of (33) was subsequently undertaken to establish the mechanism of formation of N-5-formyl-H_4-folate [73, 74]. Hydrolysis above pH 6 of the formamidinium salt **29** followed the same rate law as that for 5,10-methenyl-H_4-folate. The products of the hydrolysis were shown to be mixtures containing mostly the N-10-formyl analogue with varying amounts of the N-5-formyl isomer, the distribution depending on the pK_a of the participating buffer catalyst. At pH values less than 6, complex kinetics were observed as a consequence of isomerization of the initially formed N-10-formyl to the more stable N-1-formyl species. These results were rationalized through one kinetic scheme that relates the formamidinium salt and two isomeric N-formyl compounds to a common tetrahedral addition intermediate (37). Catalytic constants (pertaining to the term first-order in base) for various buffers were compared with the help

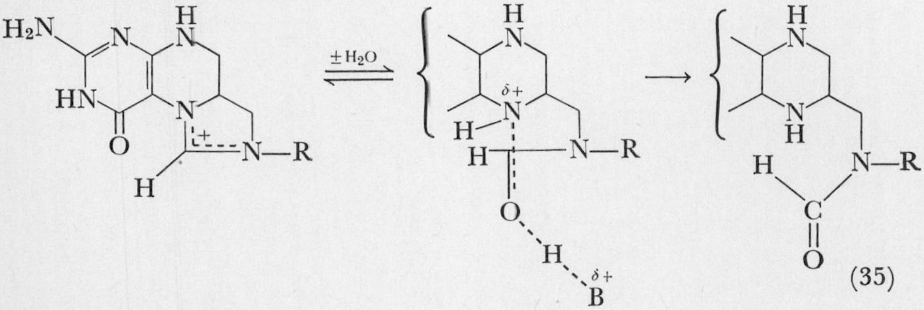

(35)

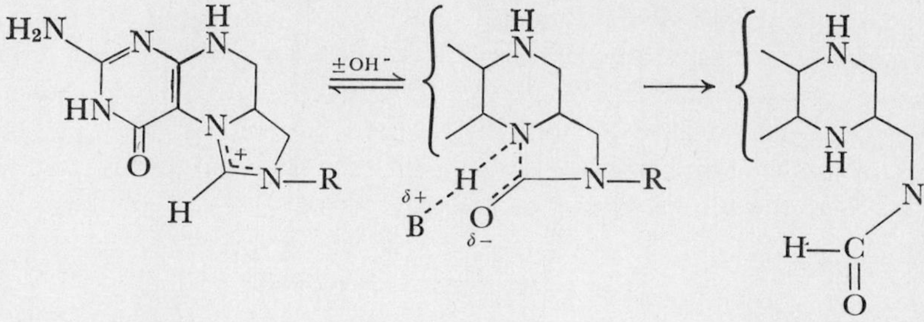

$$\qquad\qquad\qquad\qquad\qquad\qquad\qquad (36)$$

of the Brønsted relationship. Those governing formation of the N-10-formyl analogue are correlated by a straight line ($\alpha = 0.39$) in keeping with the mechanism of (35). The constants for formation of the N-1-formyl analogue, however, describe a biphasic Brønsted plot with a change from $\alpha = 1$ to $\alpha = 0$ occurring at a pK_a of about 2. The latter is interpreted in terms of a mechanism featuring protonation of the N-10 nitrogen as the rate-limiting step, the inflection and plateau ($\alpha = 0$) representing attainment of the diffusion-controlled limit.

Three points of possible interest arise from these results. (1) The equilibrium constants interrelating the formamidinium salt and the two N-formyl model compounds were found to be in nearly quantitative agreement with those for the natural cofactors, indicating the relative ground-states to be about the same and, moreover, to be independent of the remote functionality present in the latter. (2) The concurrent but unequal operation of the two distinct mechanisms in the model suggests that nonenzymic as well as enzyme catalyzed hydrolysis of the 5,10-methenyl-H_4-folate are directed toward formation of N-10-formyl-H_4-folate, which is an active formyl donor, rather than toward N-5-formyl-H_4-folate, which is not active in biological formylations. (3) Catalysis of isomerization of the N-10 isomer to N-5-formyl-H_4-folate provides a rationale for the fact that the latter is present in many tissues, but a protein exhibiting significant N-5-formyl-H_4-folate synthetase activity has not been detected [1].

Nonenzymic interconversion of the formate-carrying cofactors is by far the most well-understood of the folate reactions, and to the extent that the

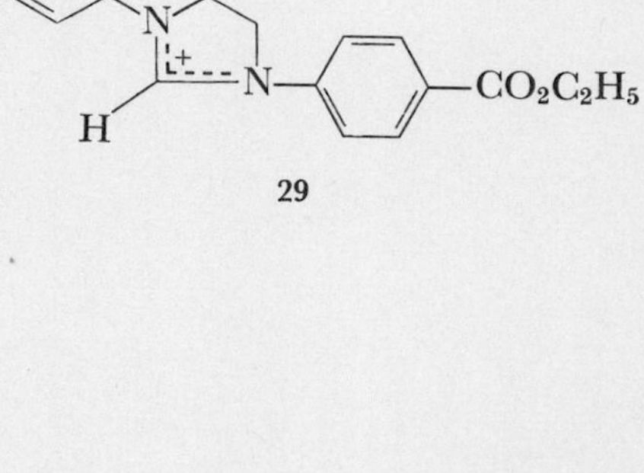

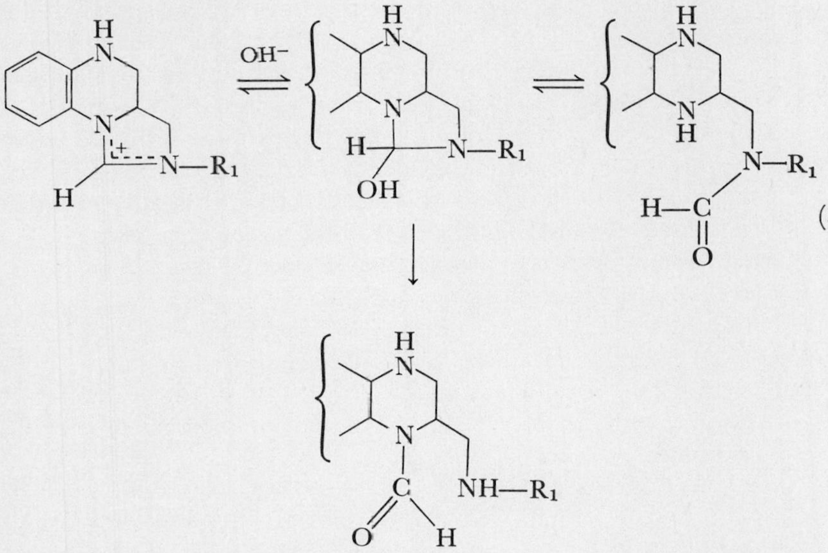

$$(37)$$

nonenzymic reactions are of biological significance these researches contribute to understanding of the flux of the cofactors in a biological medium. Extrapolation of these data to the analogous enzymic reactions is rife with difficulty, however, because of the ATP corequirement in two of the three reactions. Discussion here is limited to the 5,10-methenyl-H_4-folate cyclohydrolase.

An enzyme, 5,10-methenyltetrahydrolfolatehydrolase (decyclizing), which catalyzes the reversible hydrolysis of 5,10-methenyl-H_4-folate to N-10-formyl-H_4-folate has been purified by Lombrozo and Greenberg about 100-fold from beef liver [75]. Assuming that catalysis is effected by optimizing general acid-base interractions between the substrate and enzymic functional groups, one can envision a transition state of the type depicted in **30**, which combines the two modes of catalysis described in (35)

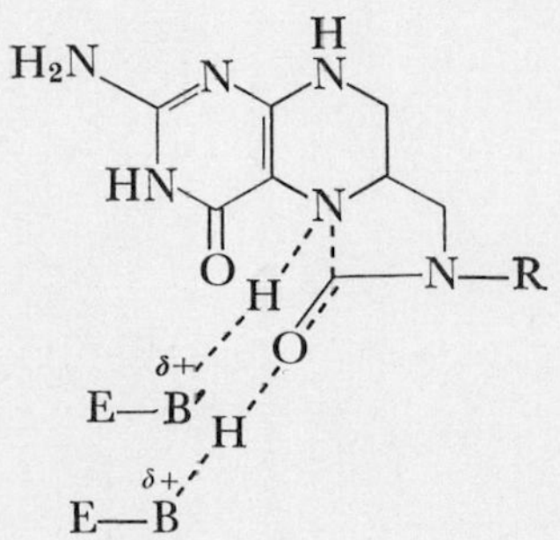

30

and (36). This mechanism might be compared with that proposed for 5-formimino cyclodeaminase (cf. **22**), since the latter exhibits cyclohydrolase activity [76]. There is no apparent difference in the nature of the catalytic residues required for the two reactions except in their respective protonation states. Because catalysis by cyclohydrolase is observed in the reverse reaction as well as the forward reaction the enzyme must be capable of assuming the inverse ionization state with respect to B and B′ of **30**. It is this ionization state that would be required for cyclodeaminase activity in terms of **22**. The question remains open as to whether two distinct proteins are involved.

In addition to the interconversions of (33) there are two important enzyme-catalyzed reactions which utilize N-10-formyl-H$_4$-folate as a donor and effect formylation of an amino function, as indicated in equations 38 and 39. In general, the requirement for N-10-formyl-H$_4$-folate as the formyl donor is in keeping with the observation that this isomer undergoes hydrolytic deformylation much more readily than does the N-5-formyl isomer [71].

The reaction of (38) is catalyzed by the enzyme, 10-formyltetrahydro-folate: 5-amino-1-ribosyl-4-imidazole-carboxamide-5′-phosphate transformy-lase, and like that of (27), lies on the pathway for biosynthesis of the purine nucleotides. Buchanan and co-workers purified the enzyme about 40-fold from extracts of acetone-dried chicken liver and reported a simultaneous 120-fold purification of inosinicase that catalyzes cyclization of the formy-lated product to give inosinic acid monophosphate (IMP) (40) [68]. Here

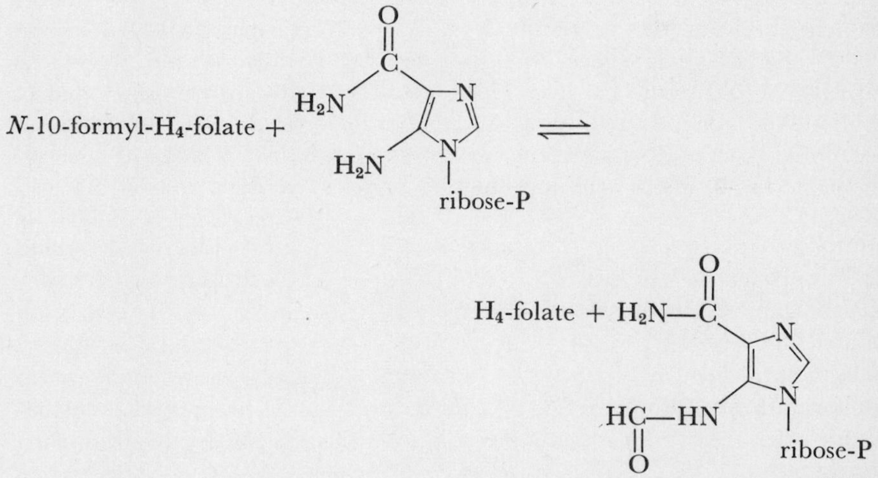

$$N\text{-10-formyl-H}_4\text{-folate} +$$

$$N\text{-10-formyl-H}_4\text{-folate} + \text{Met-tRNA} \rightleftharpoons \text{H}_4\text{-folate} + N\text{-formyl-Met-tRNA}$$

(39)

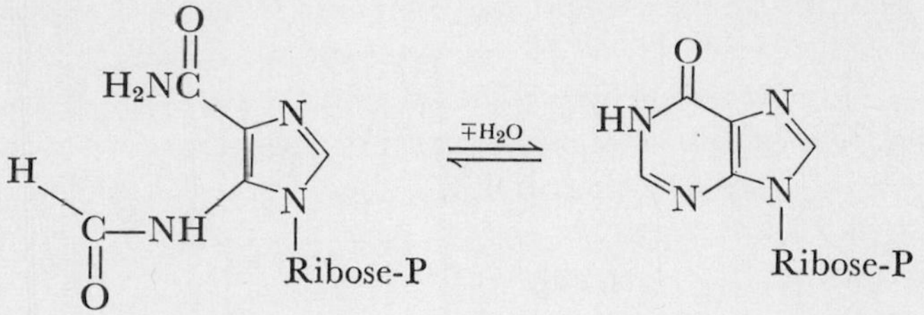

$$
\text{(40)}
$$

again, the ambiguity arises as to whether formylation and cyclization were functions of the same or of two distinct enzymes. More recent complimentation studies on *Salmonella typhimurium* [77] show, however, that transformylase and inosinicase activity can be separated into distinct proteins, thereby supporting a step-wise pathway for IMP formation. This result argues against participation of the 4(5)-carboxamido substituent in transfer of the formyl group, but does not rule out the possibility that the enzyme participates nucleophilically as suggested earlier for the mechanism of (27).

Both the foregoing transformylation and that leading to *N*-formylmet-*t*RNA may be formulated in terms of (41). Because aminolysis of formanilides is not observed in general in aqueous media due to competing hydrolysis [78], the enzymes are required to effect condensation. Data related to (37) demonstrate that aminolysis of formanilides proceeds through a tetrahedral intermediate **31** but that expulsion of hydroxide ion from the intermediate may occur with greater facility under certain conditions, that is, acidic pH, than does expulsion of an amine group. This raises the possibility that the formyl transfer proceeds through the formamidinium salt **32** through a sequence involving nucleophilic participation by the enzyme rather than direct breakdown of the tetrahedral intermediate **31**.

From the discussions concerning formyl transfer an important concept evolves with regard to the teleology implicit in the design of the natural cofactor. As in the 5,10-methylene-H_4-folate reactions of the preceding section, an argument for the importance of ΔpK_a between the *N*-5- and *N*-10 nitrogens can be advanced for formate metabolism, commencing with the substantial evidence supporting the duality of mechanism for hydrolysis of 5,10-methenyl-H_4-folate in formation of the two isomeric *N*-formyl compounds. From these and other results the following generalizations can be formulated. Although the *N*-5 derivatives—amidine, formyl or ortho-amide—are thermodynamically favored over their respective *N*-10 isomers, the latter may be kinetically accessible. Furthermore, one is tempted to speculate that hydrolysis or decomposition of the *N*-10 derivatives proceeding with expulsion of the *N*-10 nitrogen might involve proton transfer to that nitrogen in the rate-limiting step, as has been demonstrated for the model

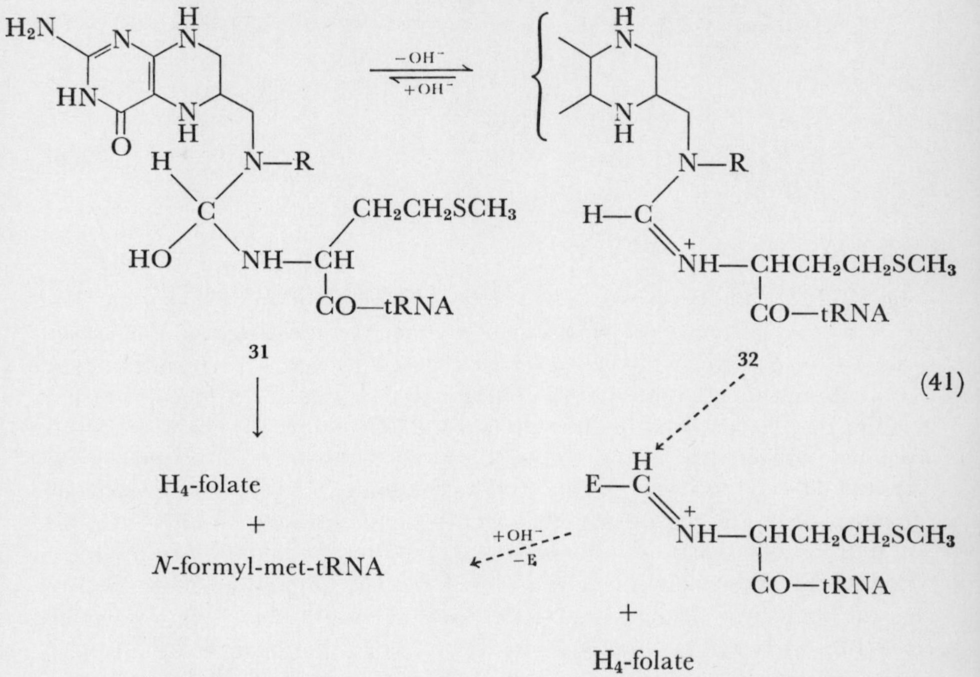

$$
\begin{array}{cc}
\textbf{31} & \textbf{32} \qquad\qquad (41)
\end{array}
$$

(Structure 31) ⇌ (−OH⁻ / +OH⁻) (Structure 32)

H_2N ... N, N, HN, O, pteridine ring with N—H; C—HO, NH—CH; $CH_2CH_2SCH_3$; CO—tRNA

Structure 32: H—C ... $\overset{+}{N}H$—$CHCH_2CH_2SCH_3$; CO—tRNA; N—R

31 →

H_4-folate

+

N-formyl-met-tRNA

+OH⁻ / −E

E—C ... $\overset{+}{N}H$—$CHCH_2CH_2SCH_3$; CO—tRNA

+

H_4-folate

transformylation (N-10-formyl → N-1-formyl). Overall, this sequence would afford a facile pathway for formate transfer that is amenable to even the simplest modes of general acid catalysis. Inspection of the reactions effecting transfer of formaldehyde or formate derived single carbon units *to* H_4-folate readily provides an argument for the importance of the more nucleophilic (more basic) N-5-nitrogen.

4.3 Activation of Formate and Formyl-H_4-folate

Three enzymic reactions (42–44) have been described that involve formation of 5,10-methenyl-H_4-folate or N-10-formyl-H_4-folate (the "active formyl" donors) coupled with cleavage of the γ phosphate from ATP. The "synthetase" reaction of (42) has been the subject of many investigations. Although the mechanism remains vague, it seems quite likely that "activation of formate" involves a step in which some oxyanion derived from formate nucleophilically displaces ADP from the γ phosphate of ATP. Because a similar step can be envisioned for the reactions of (43) and (44), discussion of the three reactions is centered around this concept, but remains hypothetical.

$$H_4\text{-folate} + HCOOH + ATP \overset{Mg^{2+}}{\rightleftharpoons} N\text{-}10\text{-formyl-}H_4\text{-folate} + ADP$$
$$+ P_i$$
$$(42)$$

$$N\text{-}5\text{-formyl-}H_4\text{-folate} + ATP \overset{Mg^{2+}}{\rightleftharpoons} 5,10\text{-methenyl-}H_4\text{-folate} + ADP$$
$$+ P_i$$
$$(43)$$

$$N\text{-}5\text{-formyl-}H_4\text{-folate} + H_2O + ATP \overset{Mg^{2+}}{\rightleftharpoons} N\text{-}10\text{-formyl-}H_4\text{-folate} + ADP$$
$$+ P_i$$
$$(44)$$

The N-10-formyl-tetrahydrofolate synthetase (EC 6.3.4.3) has been isolated from a number of sources, but the enzymes for $C.$ $cylindrosporum$ and $C.$ $acidi\text{-}urici$ have been obtained in crystalline form and have therefore received more attention than their congeners [79]. Both kinetic and exchange studies indicate that binding of substrates occurs in a random fashion but that all are bound before any products are released [80]. Although this would argue in favor of a "concerted" mechanism involving only metastable intermediates, nmr relaxation techniques (Mg^{2+} is replaced with Mn^{2+}) have provided data suggesting that both H_4-folate and N-10-formyl-H_4-folate induce conformational changes in the protein that affect the Mn^{2+}-nucleotide binding site and are necessary for activity [81]. In this context a conformational change induced by H_4-folate offers a rationale for the failure to observe any of a number of possible half reactions, for example, ATP-P_i exchange in the presence of formate [82], and allows the consideration of possible stepwise rather than concerted mechanisms.

The observation that an ^{18}O label in formate is transferred to inorganic phosphate during the course of the process has led to the hypothesis that some mixed anhydride of formic and phosphoric acids be intermediate in the reaction [82]. On this basis one is encouraged to consider only those mechanisms that involve a formylphosphate either as a discrete intermediate or as a "feature" of a concerted process, and discussion here is limited to such mechanisms. Within the formylphosphate hypothesis, one can formulate essentially two mechanisms with their attendant variations that are consistent with the data thus far reported. The mechanisms differ only in the mode of generation and nature of the formylphosphate. As a starting point the "concerted" mechanism proposed by Rabinowitz and his colleagues is depicted in (45) [80]. The aforementioned H_4-folate induced conformational change obscures the search for an experimental distinction between the "concerted" pathway and one featuring formylphosphate as a discrete enzyme-bound intermediate arising from direct nucleophilic attack of formate on the γ phosphate of ATP. Indeed, the latter is supported by a similar finding in the ATP dependent conversion of glutamate to glutamine [83]. Because the formyl synthetase exhibits no ATPase activity in the presence of formate, one must argue that the enzyme conformation in the absence of H_4-folate renders the $Mg^{2+} \cdots ATP$ complex unreactive toward formate. Binding of H_4-folate leads to the reactive conformation, but hydrolysis of the formylphosphate is precluded by the more rapid formylation of H_4-folate. It should be noted that if nucleophilicity alone governed the site of formylation, the N-5-formyl isomer would predominate.

An alternate mechanism features nucleophilic catalysis involving H_4-folate in the activation of formate. Although there is some evidence from the studies of Jaenicke and Brode, and Whiteley and Hunnekens [84, 85] for the presence of a phosphorylated H_4-folate intermediate in the formate activation catalyzed by enzymes from pigeon liver and *M. aerogenes*, respectively, these results are less than definitive owing to the impurity of the preparations. A model study predicated on this hypothesis, however, indicates such a sequence to be both thermodynamically and kinetically feasible.

Hydrolysis of the cyclic phosphoramidate **33** in formate buffer gives rise to the formanilide **35**, presumably via the hypothetical formylphosphate adduct **34** indicated [80]. It has been shown that the first-order rate constants for disappearance of **33** and formation of **35** and inorganic phosphate are the same [86] and that the formanilide does not arise from the slower reaction of aniline with formate [87]. The formolysis reaction occurs with the cyclic phosphorodiamidate but not with the acyclic species [86, 88, 89, 89a]. Recasting the reaction in the context of enzymic mediation results in a mechanism such as that of (47). In contrast to the concerted mechanism, that of (47) invokes nucleophilic catalysis by H_4-folate and a covalently bound (to substrate) formylphosphate. Note that the mechanism provides for formylation of the least nucleophilic nitrogen (N-10).

Inspection of the above mechanism involving a covalently bound formylphosphate reveals that one mole of ^{18}O (from exogeneous $H_2^{18}O$) should be incorporated into the inorganic phosphate as a consequence of phosphoramidate hydrolysis subsequent to formyl transfer. This assumes rapid exchange with medium of the H_2O formed, in cyclization of the N-5-phosphoryl-H_4-folate. Mechanism (45) does not lead to such a prediction, suggesting a possible means of differentiation. In a recent experiment no ^{18}O from solvent was incorporated into P_i for the reaction catalyzed by the synthetase from *C. cylindrosporum*. [89b].

An enzyme that catalyzes the ATP-dependent conversion of N-5-formyl-H_4-folate to 5,10-methenyl-H_4-folate (43) has been purified 400-fold from

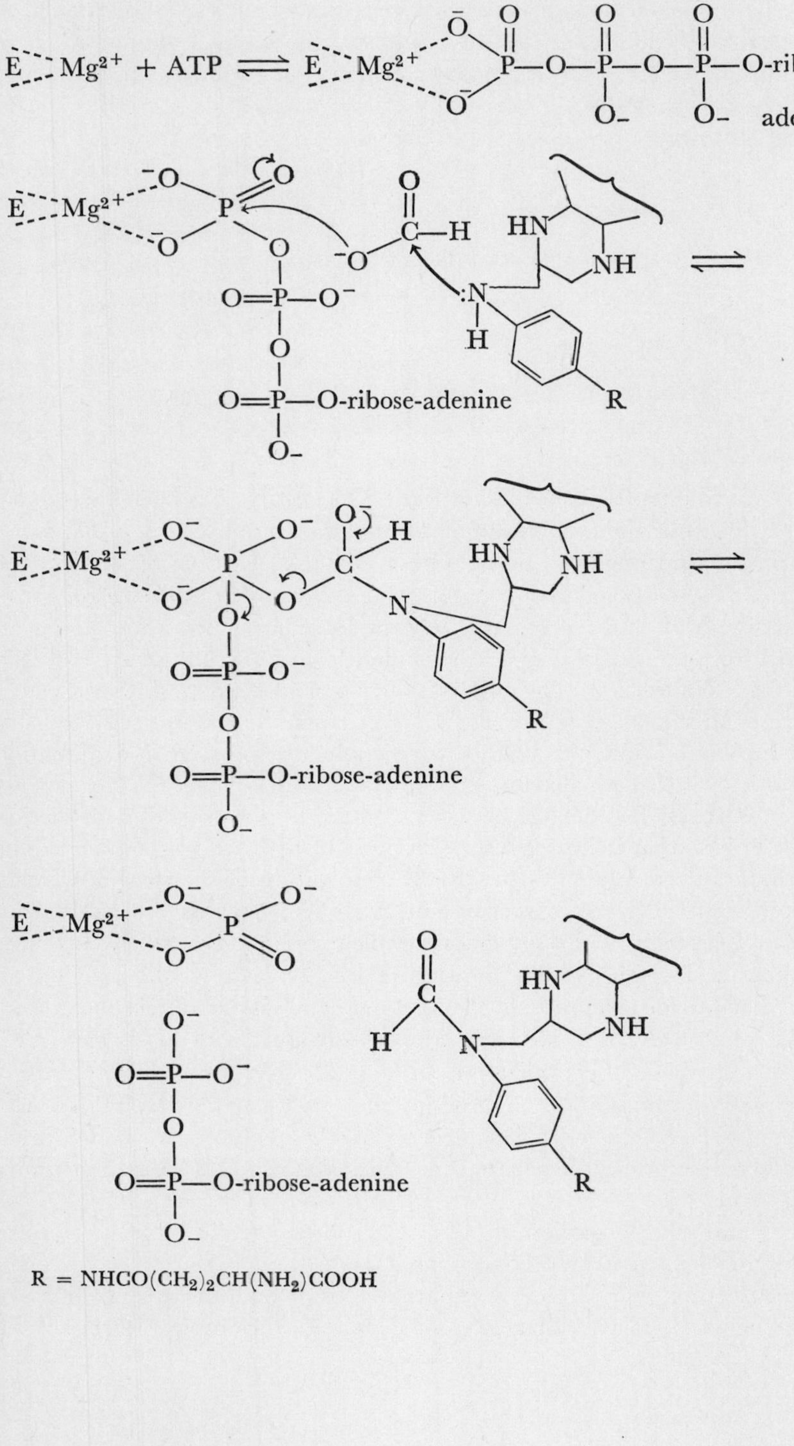

$$E \cdots Mg^{2+} + ATP \rightleftharpoons E \cdots Mg^{2+} \cdots O-P-O-P-O-P-O\text{-ribose}$$

adenine

$$\rightleftharpoons$$

$$\tag{45} \rightleftharpoons$$

O=P—O-ribose-adenine

O=P—O-ribose-adenine

O=P—O-ribose-adenine

$$R = NHCO(CH_2)_2CH(NH_2)COOH$$

167

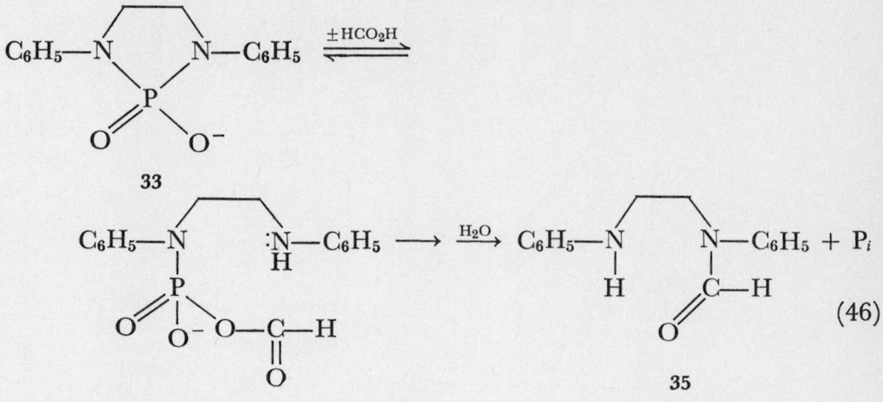

$$C_6H_5-N \qquad N-C_6H_5 \quad \overset{\pm HCO_2H}{\rightleftharpoons}$$

33

$$C_6H_5-N \qquad :\underset{H}{N}-C_6H_5 \longrightarrow \overset{H_2O}{\longrightarrow} C_6H_5-N \qquad N-C_6H_5 + P_i$$

34

35

(46)

extracts of acetone-dried sheep liver [90]. The enzyme has been the subject of only a few investigations, yielding limited data upon which a discussion of the mechanism might be based. Other enzymic reactions that effect the activation of a carboxyl group or bicarbonate coupled with hydrolysis of ATP to give ADP and inorganic phosphate have been shown to transfer an ^{18}O label from the carboxyl or bicarbonate to inorganic phosphate [83, 91], suggesting, as in the N-10-formyl-H_4-folate synthetase reaction, that a mixed anhydride intermediate is formed. Extrapolation of this concept to N-5-formyl-H_4-folate cyclodehydrase is tenable when considered in conjunction with the relevant tetrahydroquinoxaline model study of this transformation. The model reaction indicates that the rate-limiting step of the cyclization (where the reaction is not diffusion limited) involves deprotonation of the zwitterionic intermediate **37**, the extent of solvation of the oxyanion being uncertain. It follows that a species similar to **37** is formed reversibly from N-5-formyl-H_4-folate and is sufficiently stable to exist at least as a steady-state intermediate. Phosphorylation of the anionic oxygen would provide an attractive route for formation of the 5,10-methenyl-H_4-folate via the adduct **38**. The phosphorylation step provides an obvious mode by which ATP cleavage can drive the cyclization in that **38** may be so appropriately disposed for decomposition to product that its formation would be rate limiting. The mechanism predicts transfer of ^{18}O label from the N-5-formyl-H_4-folate to inorganic phosphate, and this proposition is currently under investigation.

The similar transformation of (44) was reported by Kay and co-workers to occur in chicken liver preparations [92]. Although the reaction is not well documented, the evidence apparently rules out the intermediacy of 5,10-methenyl-H_4-folate. Blakely has proposed the possibility that the

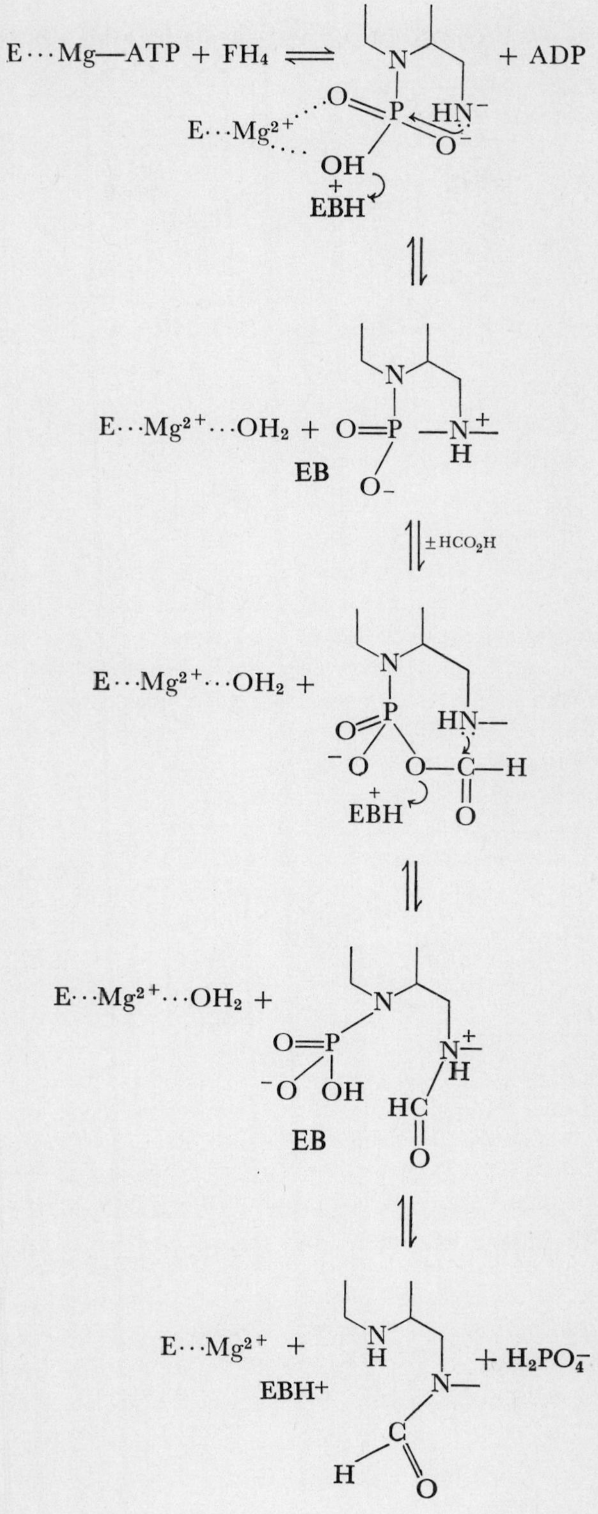

$$\text{E}\cdots\text{Mg—ATP} + \text{FH}_4 \rightleftharpoons \quad + \text{ADP}$$

$$\text{E}\cdots\text{Mg}^{2+}\cdots\text{OH}_2 + \text{EB}$$

$$\text{E}\cdots\text{Mg}^{2+}\cdots\text{OH}_2 + \quad\quad\quad\quad\quad (47)$$

$$\text{E}\cdots\text{Mg}^{2+}\cdots\text{OH}_2 + \text{EB}$$

$$\text{E}\cdots\text{Mg}^{2+} + \text{EBH}^+ \quad + \text{H}_2\text{PO}_4^-$$

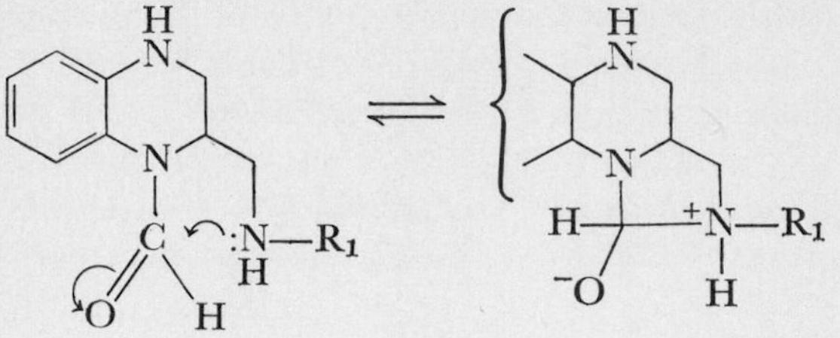

(48)

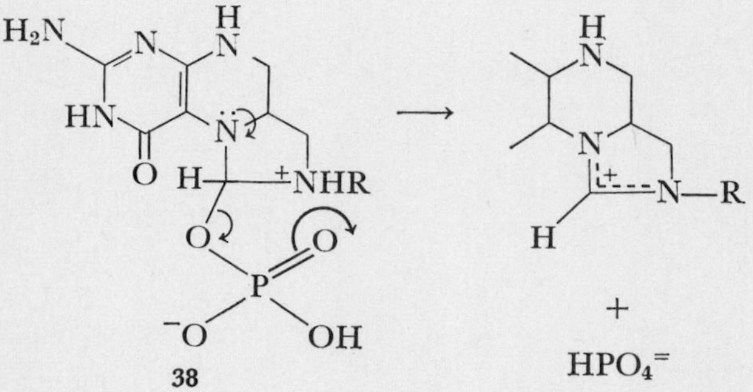

(49)

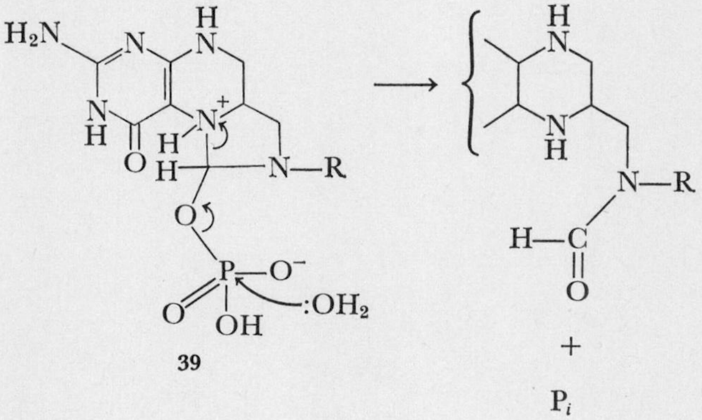

(50)

observed reaction might be a consequence of the consecutive actions of a deacylase and N-10-formyl synthetase [1]. The ambiguity presumably could be resolved by demonstrating incorporation of exogenous labeled formate into N-10-formyl-H_4-folate.

If the transformation is indeed the result of a single enzyme catalyzed reaction, a trivial modification of the proposed mechanism of N-5-formyl-H_4-folate cyclodehydrase might pertain. In the latter reaction, one might imagine decomposition of **38** with expulsion of phosphate to be greatly facilitated by participation of the nonbonding electrons of the N-5-nitrogen.

Protonation of that nitrogen (and deprotonation of the N-10 nitrogen) then would yield the phosphorylated intermediate **39** that has been stabilized with respect to phosphate expulsion. Hydrolysis of **39** via attack of water would now lead directly to the N-10-formyl product. Unlike that for the cyclode-hydrase reaction, this mechanism predicts that the ^{18}O labeled formyl would be retained and that inorganic phosphate would incorporate an oxygen atom from the water.

5 METHYL TRANSFER

The biosynthesis of methionine from homocysteine is catalyzed by two enzymes in *E. coli*. One of these proteins, 5-methyltetrahydropteroyltrigluta-mate-homocysteine transmethylase, transfers the methyl group from N-5 of the tetrahydropteroyltriglutamate to homocysteine in the presence of Mg^{+2} and phosphate ions; the second, 5-methyltetrahydrofolate-homocysteine transmethylase, contains vitamin B_{12} and catalyzes methyl transfer from N-5 of either the above pteroyltriglutamate or H_4-folate in the presence of catalytic levels of S-adenosylmethionine and a reducing system.

Taylor and Weissbach have proposed a mechanism for the latter enzyme that features premethylation of the oxidized nonfunctional cobalamin (Co^{+3}) by S-adenosylmethionine in the presence, typically of reduced flavin mononucleotide, to form an enzyme bound methyl-B_{12} [93].

$$E-Co^{+3} + CH_3-\overset{+}{\underset{|}{S}}-adenosine \xrightarrow[\text{system}]{\text{reducing}} E-Co-CH_3 \qquad (51)$$
$$\text{methionine}$$

Once methylated, the enzyme can oscillate between methylated and de-methylated reduced forms transferring methyl from N-5-methyl-H_4-folate to homocysteine without any further need for S-adenosylmethionine. It is obligatory that reducing conditions be maintained to protect the nucleo-philic $E-Co^{+1}$ formed in the cycle [94, 95]. It seems likely that methylco-balamin possesses the necessary lability to be a kinetically competent

$$E-Co-CH_3 + homocysteine \longrightarrow E-Co^{+1} + methionine \qquad (52)$$

$$E-Co^{+1} + N\text{-}5\text{-methyl-}H_4\text{-folate} \longrightarrow E-Co-CH_3 + H_4\text{-folate} \qquad (53)$$

intermediate [96]. A similar mechanism apparently applies to the transferase from porcine kidney [97].

Against this brief background, let us focus on the problem of methyl transfer to $B_{12s}(Co^{+1})$. Although the sulfonium salt, S-adenosyl methionine, readily reacts nonenzymically with B_{12s} to yield methyl-B_{12}, no nonenzymic reaction of B_{12s} with N-5-methyl-H_4-folate has been detected [1, 98]. Attempts to determine whether enzymic activation of the methyl group for transfer occurred via an oxidative pathway were unsuccessful because N-5-methyl-H_4-folate labeled at H-6 and H-7 did not release isotope from these positions during the course of the reaction. The catalytic mechanism

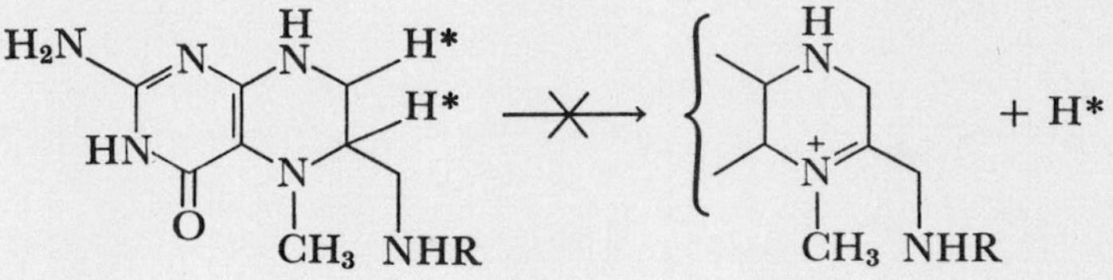

$$\tag{54}$$

40   41

for the non-B_{12} transmethylase has scarcely been studied [99]. The formation of methionine is dependent on inorganic phosphate and requires the 5-methyltetrahydropteroyltriglutamate as a methyl donor. Neither N-5-methyl-H_4-folate nor the diglutamate will substitute. Model studies of methyl transfer from quaternary N-alkyl cations to thiols in neutral or weakly basic solution did not demonstrate significant conversion to the corresponding thiol-methyl ethers except at reflux conditions [100]. For example, the N-methyl derivative of tetrahydro-8-hydroxyquinoline reacts with thiophenol to yield only 7 % of the corresponding methyl ether at 170° over a 24-hr period.

Intuitively the enzymic methyl transfer reactions must involve electrophilic activation of N-5, perhaps by protonation, followed by nucleophilic displacement. Design and characterization of appropriate model reactions remains a potentially instructive area for future endeavor.

REFERENCES

1. R. L. Blakley, *The Biochemistry of Folic Acid and Related Pteridines*, American Elsevier, New York 1969.
2. B. R. Baker, *Design of Active-Site-Directed Irreversible Enzyme Inhibitors; The Organic Chemistry of the Enzymic Active-Site*, Wiley, New York, 1967.
3. B. R. Baker, *Acc. Chem. Res.*, **2**, 129 (1969).
4. K. C. Scrimgeour and K. Smith Vitols, *Biochemistry*, **5**, 1438 (1966).
5. B. L. Hillcoat and R. L. Blakley, *Biochem. Biophys. Res. Commun.*, **15**, 303 (1964).
6. E. J. Pastore, M. Friedkin, and O. Jardetsky, *J. Am. Chem. Soc.*, **85**, 3058 (1963).
7. S. F. Zakrzewski, *J. Biol. Chem.*, **241**, 2962 (1966).
8. E. J. Pastore and K. L. Williamson, *Fed. Proc.*, **27**, 3075 (1968).
9. E. J. Pastore and O. Jardetzky, *Fed. Proc.*, **25**, 278 (1966).
10. S. F. Zakrzewski and A. Sansone, *J. Biol. Chem.*, **242**, 5661 (1967).

11. D. J. Vonderschmitt, K. Smith Vitols, F. M. Huennekens, and K. G. Scrimgeour, *Arch. Biochem. Biophys.*, **122,** 488 (1967).
12. For recent comments on the mechanism of action of nicotinamide cofactors see G. A. Hamilton, in *Progress in Bioorganic Chemistry*, Vol. I, E. T. Kaiser and F. J. Kézdy, (Eds.), Wiley, New York, 1971.
13. M. C. Archer and K. G. Scrimgeour, *Can. J. Biochem.*, **48,** 278 (1970).
14. P. Hemmerich, *Pteridine Chemistry*, W. Pfleiderer and E. C. Taylor, (Eds.), Pergamon, London, England, 1964.
15. S. Kaufman, *J. Biol. Chem.*, **239,** 332 (1964).
16. B. V. Ramasastri and R. L. Blakley, *J. Biol. Chem.*, **239,** 112 (1964).
17a. J. F. Beillmann and F. Schuber, *Biochem. Biophys. Res. Commun.*, **27,** 517 (1967).
17b. S. J. Benkovic and P. Farina, unpublished results.
18. R. E. Cathou and J. M. Buchanan, *J. Biol. Chem.*, **238,** 1746 (1963).
19. R. L. Kisliuk, *J. Biol. Chem.*, **238,** 397 (1963).
20. R. L. Blakley, *Biochem. J.*, **58,** 448 (1954).
21. R. G. Kallen and W. P. Jencks, *J. Biol. Chem.*, **241,** 5851 (1966).
22. M. J. Osborn and F. M. Huennekens, *Biochim. Biophys. Acta*, **26,** 646 (1957).
23. M. J. Osborn, P. T. Talbert, and F. M. Huennekens, *J. Am. Chem. Soc.*, **82,** 4921 (1960).
24. W. P. Jencks, *Prog. Phys. Org. Chem.*, **2,** 63 (1964).
25. S. J. Benkovic, P. A. Benkovic, and D. R. Comfort, *J. Am. Chem. Soc.*, **91,** 5270 (1969).
26. R. G. Kallen and W. P. Jencks, *J. Biol. Chem.*, **241,** 5845 (1966).
27. S. J. Benkovic, P. A. Benkovic, and R. Chrzanowski, *J. Am. Chem. Soc.*, **92,** 523 (1970).
28. V. S. Gupta and F. M. Huennekens, *Arch. Biochem. Biophys.*, **120,** 712 (1967).
29. R. G. Kallen and W. P. Jencks, *J. Biol. Chem.*, **241,** 5864 (1966).
30. L. Jaenicke, "The Mechanism of Action of Water Soluble Vitamins," *Ciba Foundation Study Groups No. 11*, A. V. S. DeRauck and M. O'Connor (Eds.), Little, Brown and Company, Boston, 1961, p. 38.
31. F. M. Huennekens, H. R. Whiteley, and M. J. Osborn, *J. Cell. Comp. Physiol.*, **54,** 109 (1959).
32. E. C. Wagner, *J. Org. Chem.*, **19,** 1862 (1954).
33. H. Volz and H. H. Kiltz, *Tetrahedron Lett.*, **22,** 1917 (1970).
34. J. DeLuis, doctoral thesis, The Pennsylvania State University, 1964.
35. P. Reyes and C. Heidelberger, *Mol. Pharm.*, **1,** 14 (1965).
36. M. Y. Lorenson, G. F. Maley, and F. Maley, *J. Biol. Chem.*, **242,** 3332 (1967).
37. M. I. S. Lomax and G. R. Greenberg, *J. Biol. Chem.*, **242,** 1302 (1967).
38. J. G. Flaks and S. S. Cohen, *J. Biol. Chem.*, **234,** 2981 (1959).
39. D. V. Santi and T. T. Sakai, *Biochem. Biophys. Res. Comm.*, **42,** 813 (1971).
40. T. C. Crusberg, R. Leary, and R. L. Kisliuk, *J. Biol. Chem.*, **245,** 5292 (1970).
41. A. Fridland and C. Heidelberger, *Fed. Proc.*, **29,** 878 (1970).
42. R. B. Dunlap, N. G. L. Harding, and F. M. Huennekens, *Biochemistry*, **10,** 88 (1971).
43. R. L. Blakley, B. V. Ramasastri, and B. M. McDougall, *J. Biol. Chem.*, **238,** 3075 (1963).
44. E. J. Pastore and M. Friedkin, *J. Biol. Chem.*, **237,** 3802 (1962).
45. S. R. Heller, *Biochem. Biophys. Res. Commun.*, **32,** 998 (1968).
46. D. V. Santi, C. F. Brewer, and D. Farber, *J. Heterocycl. Chem.*, **7,** 903 (1970).
47. T. I. Kalman, *Biochemistry*, **10,** 2567 (1971).
48. D. V. Santi and T. T. Sakai, *Biochemistry*, **10,** 3598 (1971).
49. M. Friedkin and A. Kornberg in *The Chemical Bases of Heredity*, W. D. McElroy and B. Glass (Eds.), Johns Hopkins Press, Baltimore, 1957, p. 609.

50. V. S. Gupta and F. M. Huennekens, *Biochemistry*, **6,** 2168 (1967).
51. D. V. Santi and A. L. Pogolotti, Jr., *J. Heterocycl. Chem.*, **8,** 265 (1971).
52. W. Wilmanns, B. Rücker, and L. Jaenicke, *Z. Physiol. Chem.*, **322,** 283 (1960).
53. R. B. Woodward and R. Hoffmann, *The Conservation of Orbital Symmetry*, Verlag Chemie, Weinheim, Germany, 1970.
54. N. J. Leonard and A. S. Hay, *J. Am. Chem. Soc.*, **78,** 1984 (1956).
55. A. J. Wahba and M. Friedkin, *J. Biol. Chem.*, **237,** 3794 (1962).
56. Y. C. Yeh and G. R. Greenberg, *J. Biol. Chem.*, **242,** 1307 (1967).
57. L. Schirch and A. Diller, *J. Biol. Chem.*, **246,** 3961 (1971) and references therein.
58. L. Schirch and W. T. Jenkins, *J. Biol. Chem.*, **239,** 3801 (1964).
59. P. M. Jordan and M. Akhtar, *Biochem. J.*, **116,** 277 (1970).
59a. P. Besmer and D. Arigoni, personal communication.
60. T. C. Bruice and S. J. Benkovic, *Bioorganic Mechanisms, Vol. II*, Benjamin, New York, 1967.
61. R. H. DeWolfe, *Carboxylic Ortho Acid Derivatives*, Academic Press, New York, 1970.
62. S. J. Benkovic, T. H. Barrows, and W. P. Bullard, unpublished results.
63. W. P. Jencks, "Catalysis in Chemistry and Enzymology," McGraw-Hill, New York, 1969.
64. A. Miller and H. Waelsch, *J. Biol. Chem.*, **228,** 397 (1957).
65. J. C. Rabinowitz in *The Enzymes*, Vol. 2, 2nd ed., P. D. Boyer, H. Lardy, and K. Myrbach (Eds.), Academic Press, New York, 1960, p. 185.
66. S. J. Benkovic, W. P. Bullard, and P. A. Benkovic, *J. Am. Chem. Soc.*, **94,** 7542 (1972).
67. E. Grazi and N. Rossi, *J. Biol. Chem.*, **243,** 538 (1968).
68. L. Warren and J. M. Buchanan, *J. Biol. Chem.*, **229,** 613 (1957).
69. S. J. Benkovic, W. P. Bullard, and T. H. Barrows, unpublished results.
70. D. R. Robinson and W. P. Jencks, *J. Am. Chem. Soc.*, **89,** 7098 (1967).
71. M. May, T. J. Bardos, F. L. Barger, M. Lansford, J. M. Ravel, G. L. Sutherland, and W. Shive, *J. Am. Chem. Soc.*, **73,** 3067 (1951).
72. D. R. Robinson and W. P. Jencks, *J. Am. Chem. Soc.*, **89,** 7088 (1967).
73. S. J. Benkovic, W. P. Bullard, and P. A. Benkovic, see reference 66.
74. W. P. Bullard, S. J. Benkovic, submitted to *J. Am. Chem. Soc.*
75. L. Lombrozo and D. M. Greenberg, *Arch. Biochem. Biophys.*, **118,** 297 (1967).
76. K. Uyeda and J. C. Rabinowitz, *J. Biol. Chem.*, **242,** 24 (1967).
77. F. R. Dalal and J. S. Gots, *Fed. Proc.*, 679 (1967).
78. W. P. Jencks, B. Schaffhausen, K. Tornheim, and H. White, *J. Am. Chem. Soc.*, **93,** 3917 (1971).
79. J. C. Rabinowitz and W. E. Pricer, Jr., *J. Biol. Chem.*, **237,** 2898 (1962).
80. K. Uyeda and J. C. Rabinowitz, *Arch. Biochem. Biophys.*, **107,** 419 (1964).
81. R. H. Himes and M. Cohn, *J. Biol. Chem.*, **242,** 3628 (1967).
82. R. H. Himes and J. C. Rabinowitz, *J. Biol. Chem.*, **237,** 2915 (1962).
83. M. Orlowski and A. Meister in *The Enzymes*, Vol. IV, 3rd ed., P. D. Boyer (Ed.), Academic Press, New York, 1971.
84. L. Jaenicke and G. Brode, *Biochem. Z.*, **334,** 108 (1961).
85. H. R. Whiteley and F. M. Huennekens, *J. Biol. Chem.*, **237,** 1290 (1962).
86. S. J. Benkovic and B. A. Amato, unpublished results.
87. G. Di Sabato and W. P. Jencks, *J. Am. Chem. Soc.*, **83,** 4393, 4400 (1961).
88. I. Öney and M. Caplow, *J. Am. Chem. Soc.*, **89,** 6972 (1967).
89a. C. Kutzbach and L. Jaenicke, *Ann. Chem.*, **692,** 26 (1966).
89b. W. Bullard, unpublished results.

90. D. M. Greenberg, L. K. Wynston, and A. Nagabhushanam, *Biochemistry*, **4,** 1872 (1965).
91. Y. Kaziro, L. F. Hass, P. D. Boyer and S. Ochoa, *J. Biol. Chem.*, **237,** 1460 (1962).
92. L. D. Kay, M. J. Osborn, Y. Hatefi and F. M. Huennekens, *J. Biol. Chem.*, **235,** 195 (1960).
93. R. T. Taylor and H. Weissbach, *Arch. Biochem. Biophys.*, **129,** 745 (1969).
94. R. T. Taylor and M. Hanna, *Arch. Biochem. Biophys.*, **137,** 453 (1970).
95. H. Ruediger and L. Jaenicke, *Eur. J. Biochem.*, **10,** 557 (1969).
96. R. T. Taylor, *Arch. Biochem. Biophys.*, **144,** 352 (1971).
97. G. T. Burke, J. H. Mangum, and J. D. Brodie, *Biochemistry*, **9,** 4297 (1970).
98. W. Friedrich and E. Koenigk, *Biochem. Z.*, **336,** 444 (1962).
99. C. D. Whitfield, E. J. Steers, Jr., and H. Weissbach, *J. Biol. Chem.*, **245,** 390 (1970).
100. G. N. Schrauzer and R. J. Windgassen, *J. Am. Chem. Soc.*, **89,** 3607 (1967).

CATALYTIC ACTIVITY OF PEPSIN

Department of Biological Chemistry, Hahnemann Medical School,
Philadelphia, Pennsylvania

1 INTRODUCTION

Pepsin is an endopeptidase (proteolytic enzyme) found in the gastric juice of all vertebrates, capable of hydrolyzing selectively proteins in the acid pH range of 1–6 [1]. Although it was the second enzyme to be crystallized (urease being the first) [2], only in the last 10 years has its complex mechanism of action begun to be unraveled. This delay in investigating pepsin action was primarily due to the lack of diversity in the available substrates and to the experimental difficulties inherent to the accurate measurement of peptide hydrolysis in the acidic range. Progress in the enzymology of pepsin has been discussed recently by Fruton [3], and his excellent review article should be referred to for general background information.

The present review focuses on the mechanism of action of porcine pepsin, although a number of other pepsins and pepsinlike proteases have recently received study. These include pepsins and/or pepsinogens derived from the stomach mucosa of humans [4–6], cattle [7, 8], chicken [9, 10], and fish, that is, dogfish [11]. Each source of pepsin or pepsinogen yields a multiplicity of enzymes and zymogens. All these enzymes and their precursors have been highly purified and often crystallized. They all resemble the porcine enzyme in molecular weight (about 35,000 and 40,000, respectively, for pepsin and pepsinogen), in amino acid content (all have an excess of acidic amino acids), and in their ability to cleave synthetic dipeptides. The dissimilarities, mostly in kinetic specificity, are surprisingly few, considering the widely differing sources of these enzymes.

A large number of other acidic proteinases resemble pepsin in their catalytic action. Gastricsin [12] has been isolated and purified from human gastric juice. The corresponding zymogen has not yet been identified. Several mold proteases with pepsinlike properties have also been described [13, 14].

Rennin is probably the best known pepsinlike acidic protease. Rennin obtained from the fourth calf stomach has a high milk-clotting ability and is used in cheese manufacturing. Many of the physical and chemical properties of the protein have been characterized, but there are no kinetic data available for the discussion of its mechanism of action [15]. Rennin has an N-terminal glycine, C-terminal alanine, a molecular weight of 31,000, an acid to basic amino acid content of 29/16, and a 70 % homology in its amino acid sequence with pepsin [16, 17]. The conversion of prorennin to rennin experimentally is identical to that of pepsinogen to pepsin. Rennin also reacts with diazoacetyl derivatives to give an inactive enzyme [18]. This undoubtedly indicates a participation of carboxyl group(s) in the active site of rennin. Rennin's principal function in the fourth calf stomach is to clot milk by specifically cleaving a phenylalanyl-methionine bond in κ-casein, a particularly well-studied specific cleavage [15, 19]. Crystalline rennin contains, in fact, three slightly different enzymes. Renninlike fungal proteases have also been reported [20].

2 ACTIVATION OF PEPSINOGEN-PURIFICATION OF PEPSIN

Because most of the papers dealing with the mechanism of pepsin action use crystalline porcine pepsin, this is the enzyme preparation we are concerned with in the rest of the chapter. In most studies this commercially available enzyme is used without further purification. It is probably less than 50% active enzyme. As with most proteases, pepsinogen can be obtained in a much purer state than pepsin. Pepsinogen is stable above pH 6 but is converted to pepsin at pH's below 5. This conversion of pepsinogen to pepsin at acid pH apparently involves an intramolecular cleavage of the zymogen. The first-order rate constant of this activation is independent of pepsinogen concentration, and pepsin does not increase the rate of auto-activation [21]. In a classic paper, Rajagopalan, Moore, and Stein [22], presented a detailed procedure for the activation of pepsinogen and purification of the resulting pepsin. At pH 2 complete activation of pepsinogen occurs within 20 min, and a subsequent chromatography on SE Sephadex C-25 of the activation mixture yields "pure pepsin" separated from its activation peptides. An added advantage of this method is that pepsin obtained in solution is at a convenient concentration for use in most kinetic studies [23]. The activation involves splitting off several peptides from only the N-terminal end of pepsinogen (mol wt 41,000, amino acid content 363 residues [23]) to give active pepsin (mol wt 34,163; 321 amino acid residues [22]). Of the 42 amino acid residues lost during activation, 9 are lysine, 2 are histidine, and 2 are arginine.

The single chain pepsin contains only 4 basic residues and 39 glutamic and aspartic acids. A report has recently appeared [24] that describes the use of a single long SE Sephadex C-25 column to remove basic peptides and desalt the pepsin prepared from pepsinogen. This is in contrast to the usual two-column procedure, one to remove peptides and one to desalt the solution. This latter procedure [24] can yield either pepsin frozen in solution or lyophilized. One of the principal differences in the kinetic behavior of commercial pepsin versus pepsin from pepsinogen is that the latter exhibits larger K_m values [23, 24]. If one considers the complexity of the experimental part of recent papers on pepsin, the use of these purification procedures should promote the use of purer pepsin preparations. Finally, proteinases of aspergillus oryzae activate pepsinogen at pH 5 to form "leucyl-pepsin" [25]. This leucyl-pepsin has one extra amino acid (leucine) on its N-terminal end and possesses essentially the same proteolytic activity as the normal N-terminal isoleucyl-pepsin, suggesting a nonessential role for the N-terminal group.

2.1 Nomenclature

Because of the multiplicity of pepsinogens and corresponding pepsins isolated from pig stomach at various times by different workers, the nomenclature of the pepsins is ambiguous [26]. However, we use the nomenclature of Ryle, in which the major component of commercial pepsin and pepsinogen ($> 85\%$), is called A. These A proteins, containing a mono-bonded phosphate group [27], are the ones most widely used for the mechanistic studies. The phosphate group does not seem to play any mechanistic role [27]. The minor components, pepsinogen and pepsin B and C, are similar to the A material in molecular weight and general physico-chemical properties [28–30]. The main differences arise in the activity of pepsins B and C toward hemoglobin and dipeptides in their milk-clotting ability and the absence of a phosphate residue. Pepsinogen and pepsin D appear to be identical to the dephosphorylated A material [31, 32].

2.2 Linear Sequence

Recently a large number of papers have appeared dealing with the linear amino acid sequence of both porcine pepsin and pepsinogen [33, 34]. Both pepsinogen A and pepsin A are composed of a single peptide chain with alanine at the C terminal and leucine and isoleucine, respectively, at the N-terminal position. The identical C-terminal amino acid residue suggests that activation of pepsinogen to pepsin involves only N-terminal cleavage. Pepsin (and therefore pepsinogen) contains three disulfide bridges, two of which are in small loops of five and six residues, and the other in a larger loop containing many acidic amino acids and the single histidine [35] (cf. section on inhibition by chemical modification). Four unique methionine-containing peptides were also sequenced. The single phosphate residue is on a serine in the following sequence: Glu-Ala-Thr-Ser$P$-Glu-Glu-Leu-Ser-Ile-Thr-Tyr localized in the N-terminal half of the protein chain [36]. The isolation of only one phosphopeptide fragment also supports other evidence [27] suggesting a phosphomonoester bond in pepsin.

Along with the N-terminal sequence of 9-amino acids, Ile-Gly-Asp-Glu-Pro-Leu-Glu-Asn-Tyr [37], a number of other peptide sequences in this half of the enzyme have also been determined [38, 39]. A 43-residue C-terminal peptide has recently been reported [40] with earlier work establishing shorter sequences of up to 27 residues [41, 42]. The characteristic feature of this N-terminal sequence of pepsin is that it contains one lysine and two arginines (three out of the four basic residues) [43]. Considering the amount of effort now being devoted to determining the linear amino acid sequence of pepsin, one should expect a completed sequence within a year or two. The one discrepancy arising to date in these studies concerns the number of tryptophan residues, variously reported to be four [44], five [45], or six [46]. No tryptophan residues appear to be lost from pepsinogen during activation. The sequences around four of the tryptophans have been determined [44]. The "active site" peptide (Ile-Val-Asp-Thr-Gly-Thr-Ser) has been identified by using diazo compounds to esterify the aspartic acid and produce an inactive enzyme [47]. This reaction is discussed in greater detail later in the review. An identical "active site" peptide has been reported for pepsin C (Ile-Val-Asp-Thr) [48]. Apparently the active site of porcine pepsin is located in the large C-terminal end of the molecule [49].

2.3 Pepsin Assay

Kinetic studies play an important role in determining the mechanism of enzyme action. To obtain comparable results, the "operational normality" [50] of the enzyme solution must be known. Because diazoacetyl compounds stoichiometrically react with active pepsin in a one-to-one ratio, these would seem to be ideal reagents for determining the operational normality of pepsin solutions. Stepanov [18] has shown that pepsin will react with the colored inhibitor, N-diazoacetyl-N'-2,4-dinitrophenyl-ethylenediamine (DDE) (refer to Table 8). The absorption spectrum of DDE inhibited pepsin exhibits maxima at 280 nm and 360 nm. The peak at 360 nm, should surely be useful for a spectrophotometric titration similar to that first developed for α-chymotrypsin [51]. A diazoacetyl compound that has the backbone of phenylalanine is shown below. This compound [52] should have both the spectral properties and specificity to be useful for a spectrophotometric titration of pepsin.

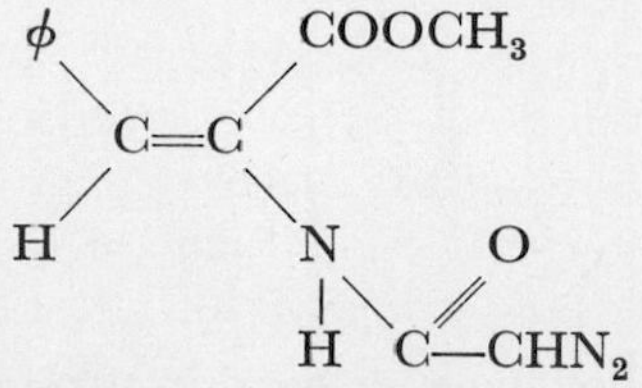

$$\phi \diagdown \qquad COOCH_3$$

$$C\!=\!C$$

$$H \diagup \qquad N \qquad O$$

$$\begin{array}{c} | \\ H \end{array} \quad \begin{array}{c} \\ C\!-\!CHN_2 \end{array}$$

In lieu of a stoichiometric titration for determining the concentration of active pepsin, a number of assay methods have been developed, together with the determination of the absorbance of pepsin solutions at 278 nm, assuming a molar absorptivity of 50,900 [53]. The oldest and most widely used rate assay is that of Anson [54], in which the rate of the pepsin catalyzed hydrolysis of acid denatured hemoglobin is measured at 280 nm at pH 1.8 and 37°. One unit of enzyme activity is defined as the amount producing $A_{280} = 0.001/min$ of trichloroacetic acid soluble products. Commercial preparations assay at 2500 units per milligram. The potential pepsin activity of pepsinogen is about 3000 units per milligram; pepsin prepared by the method of Rajagopalan et al. [22] assays at about 4000 units per milligram. The sensitivity of this method has been increased by using radioactive iodine-labeled hemoglobin [55]. Also, Anson's method has been simplified [22]. A simple and rapid nephelometric pepsin assay has recently been developed employing a fine, uniform suspension of bovine serum albumin as the substrate [56]. This method, however, appears to be most applicable to clinical situations.

A number of synthetic peptide substrates have been suggested as standards for rate assays. Acetyl-L-phenylalanyl diiodo-L-tyrosine, as one of the first substrates kinetically investigated [57, 58], has been employed by following the rate of release of the ninhydrin positive diiodo-L-tyrosine at pH 2.0. For the earlier work (before 1966) this involved the laborious and time-consuming procedure of analyzing aliquots. More recently the Technicon Auto Analyzer automated ninhydrin method [59] has been used frequently to follow the production of ninhydrin positive material [23, 60, 61]. Silver [62] has adopted the spectrophotometric method of Schwert and Takenaka [63] for following the cleavage of Ac-Phe ⤙ Tyr* by measuring the decrease in absorbance at 237 nm. However, pepsin also absorbs at this wavelength, greatly complicating the measurement of the small absorbance changes resulting from the splitting of the phenylalanyltyrosine bond. Finally, Fruton et al. [64] have synthesized a number of substrates for pepsin of the type Z-His-Phe(NO$_2$) ⤙ Phe-OMe. The major advantage of these p-nitrophenylalanine substrates is that the rate of cleavage can be followed spectrophotometrically at 310 nm, a wavelength at which pepsin exhibits only a small absorbance. Finally, it should be noted that nmr has been used for the determination of the stereospecific hydrolytic action of pepsin [65].

These rate assays yield only relative pepsin concentrations in units of moles/(liter)(sec)(mg) instead of the more useful moles/liter. Until a procedure for determining the operational normality of a pepsin solution has been developed, agreement on *one* particular rate assay as a standard would be exceedingly useful [66].

3 SPECIFICITY

The first synthetic substrates for pepsin, Z-Glu ⤙ Tyr and Z-Glu ⤙ Tyr-NH$_2$, were employed over 30 years ago [67]. Toward these dipeptides

---

* The symbol ⤙ indicates the bond cleaved during pepsin catalyzed hydrolysis.

and their ethyl esters, the enzyme exhibits a maximum activity at around pH 4 with a $k_{cat} \cong 10^{-3} sec^{-1}$ and $K_m \cong 2$ mM at 31.6° [68, 69]. These substrates are not well suited for kinetic studies because they are hydrolyzed very slowly by pepsin; they ionize in the pH region of pepsin activity (pH 1–6); and they are only sparingly soluble, especially at low pH. One or more of these disadvantages applies to almost every synthetic substrate hydrolyzed by pepsin, thereby rendering the kinetic study of pepsin action rather difficult from the experimental point of view. Baker [57, 70] made the important observation that the aromatic dipeptides Ac-Phe $\leftarrow$ Tyr, Ac-Tyr $\leftarrow$ Tyr and Ac-Phe $\leftarrow$ diiodoTyr are hydrolyzed by pepsin more rapidly than Z-Glu $\leftarrow$ Tyr.

Pepsin is one of the primary enzymes used to obtain peptides for the determination of the amino acid sequence of proteins or polypeptides. For this purpose pepsin is generally used at pH 2 where it exhibits optimal activity toward proteins that are denatured at this low pH. Hill [71] recently reviewed pepsin's specificity toward proteins. At first inspection pepsin seems to exhibit a broad specificity, probably because of the experimental conditions of high enzyme concentration for an extended time period. A closer examination indicates, with few exceptions, that pepsin prefers to hydrolyze bonds formed from aromatic amino acids or leucine. For example, in hemoglobin, 10 of 11 aromatic residues in the α-chain and 11 of 13 in the β-chain are cleaved by pepsin. When adjacent susceptible peptide bonds occur, pepsin usually cleaves only one because of the formation of an α-amino or α-carboxyl group adjacent to the less susceptible bond. A number of smaller peptides have also been synthesized for specificity studies.

There are generally two types of substrates used for pepsin; the aromatic dipeptides modeled after the work of Baker [57, 70] and the cationic substrates designed by Fruton and co-workers. The advantages and disadvantages of these substrates are as follows:

Aromatic dipeptides:

1. Medium solubility—maximum solubility approaches K_m.

2. Several different research groups have used these substrates at various temperatures and varying concentrations of different organic solvents. Direct comparison of data is difficult.

3. Some ionize in the pH region of interest, making pH dependency studies more complex.

4. The neutral substrates are the best yet devised for pH dependency studies across the entire pH range of 1–6.

Cationic peptides:

1. Large amount of data determined under identical conditions, allowing ready comparisons.

2. Excellent solubility at low pH but extremely poor solubility above pH 4. For most of these substrates, complete pH dependency studies are *not* possible due to this lack of solubility in the pH range 4–6.

3 At low pH their solubility is several times that of K_m.

4. No organic solvents have been used with these peptides, and all kinetic studies were done at 37°.

5. Many of these peptides exhibit enhanced reactivity in comparison to the aromatic dipeptides. In fact, one area that should be investigated is to use the less-reactive aromatic peptides as inhibitors during the pepsin catalyzed hydrolysis of the better cationic peptides. This would allow measurement of the K_I for these aromatic peptides and comparison to the K_m value. The significance of these measurements is discussed in the section on Inhibition.

Table 1 compares the available kinetic constants for the pepsin catalyzed hydrolysis of aromatic dipeptides. The pH 2 results were chosen for comparison because of the large number of data available at this pH. The pH dependency and inhibition studies with these compounds, if available, are discussed later. An examination of Table 1 provides the following information:

1. Phenylalanine residues are bound more strongly than tyrosine residues, suggesting that the binding site of the active center of pepsin is hydrophobic.

2. Binding of dipeptides to pepsin involves both side chains.

3. The position of the aromatic amino acid affects mainly the k_{cat} values.

4. The best of the aromatic dipeptide substrates contain a dihalogenated tyrosine residue.

5. If a covalently bound intermediate is formed during hydrolysis, the rate-limiting step is probably the formation rather than the subsequent hydrolysis of this intermediate (refer to the section on D_2O effects). A comparison of k_{cat} for compounds 2, 3, 4, and 13, 14, 15 (Table 1) shows that they react with different rate constants. This must mean that the rate-determining step for the hydrolysis of these substrates is not the breakdown of a common intermediate.

6. The lack of substrate solubility seems to be a major problem for every pepsin substrate at least at some pH values. The N-gluconyl group has been suggested as a protecting group that also acts as a solubilizing agent (cf. number 7, Table 1) [73]. Although the solubility of the dipeptide containing this group increases by a factor of 3 over the acetyl compound, the K_m also increases by a factor of 2. This essentially offsets the usefulness of this protecting group.

7. These studies have provided evidence that there are no large ionic strength effects in the range I = 0.01 to 0.1 [58, 74] nor specific salt effects

TABLE 1 KINETIC CONSTANTS FOR THE PEPSIN CATALYZED HYDROLYSIS OF AROMATIC PEPTIDES AT pH 2

Substrate	Reference	Temperature	k_{cat} sec^{-1}	$K_m \cdot 10^3$ (M)
1. Ac-Phe-diiodoTyr	58	37°	0.20	0.0750
2. Ac-Tyr-Tyr	72	37°	0.015	6.1
3. Ac-Tyr-Phe	72	37°	0.005	2.0
4. Ac-Phe-Tyr	72	37°	0.085	2.2
			$(0.07)^a$	(2.3)
5. Ac-Phe-Phe	72	37°	0.015	0.16
Ac-Phe-Phe	90	37°	0.038	1.4
6. Ac-Phe-Tyr-OMe	73	37°	0.073	1.75
7. Gluc-Phe-Tyr-OMe	73	37°	0.026	3.25
8. Ac-Phe dibromoTyr^b	74	25°	0.073	0.14
9. Z-Phe-Tyr^c	72	35°	0.0124	0.21
10. Ac-Phe-Tyr^c	72	35°	0.0466	1.95
	75	35°	0.037	1.43
11. Ac-Phe-Tyr-NH$_2$^{c}	75	35°	0.078	2.41
12. Ac-Phe-Trp^c	75	35°	0.0524	0.7
13. Ac-Phe-Tyr-OMe^d	23	25°	0.0187	2.4
			(0.029)	(3.9)
14. Ac-Phe-Phe-OMe^{d,e}	23	25°	(0.0107)	(1.4)
15. Ac-Tyr-Phe-OMe^{d,e}	23	25°	(0.00134)	(2.8)

^{a} Values in parentheses determined with pepsin prepared by the activation of pepsinogen.
^{b} 5.0% methanol.
^{c} 3.4% methanol.
^{d} 3.2% dioxane.
^{e} pH 2.3.

for buffers of acetate, formate, chloroacetate, citrate, phosphate, and hydrochloric acid [23, 76].

Although the specificity requirements of the active site of pepsin can be gleaned from the data in Table 1, these studies were primarily designed to obtain other mechanistic information. Fruton and co-workers have looked systematically at the specificity requirements of pepsin by measuring the hydrolysis of many cationic substrates. The results of these studies are found in Tables 2, 3, and 4. Since three reviews on this subject have appeared recently [3, 77, 78] only the main points of these studies are listed herein. These substrates will be classified as A — X ⤙ Y — B where X and Y are amino acids (mostly aromatic), the X ⤙ Y bond is hydrolyzed by pepsin, and A and B are the protecting groups on the N-terminal and C-terminal side, respectively.

An examination of all the kinetic data reported in Tables 2, 3, and 4 shows similarity in three areas. First, as the substrates are systematically varied

TABLE 2 SIDE CHAIN SPECIFICITY OF PEPSIN KINETICS OF THE PEPSIN CATALYZED HYDROLYSIS OF SUBSTRATES OF THE TYPE Z-His-X$\nwarrow$Y-OMe at pH 4[a]

Substrate	k_{cat} (sec^{-1})	K_m (mM)	k_{cat}/K_m (sec^{-1} mM^{-1})
1. Z-His-Gly-Phe-OMe	0.0014	1.8	0.0007
2. Z-His-Ala-Phe-OMe	0.0023	1.7	0.0014
3. Z-His-Nva-Phe-OMe	0.011	1.0	0.11
4. Z-His-Leu-Phe-OMe	0.017	0.5	0.034
5. Z-His-Phe-Phe-OMe	0.17	0.33	0.52
6. Z-His-Cha-Phe-OMe	0.015	0.17	0.088
7. Z-His-Tyr-Phe-OMe	0.009	0.30	0.030
8. Z-His-Trp-Phe-OMe	0.013	0.25	0.052
9. Z-His-Phe-Gly-OMe	0.0021	1.6	0.0013
10. Z-His-Phe-Ala-OMe	0.0037	1.8	0.0021
11. Z-His-Phe-Nle-OMe	0.007	0.4	0.018
12. Z-His-Phe-Leu-OMe	0.0052	0.5	0.010
13. Z-His-Phe-Cha-OMe	0.0026	0.3	0.009
14. Z-His-Phe-Tyr-OMe	0.17	0.29	0.59
15. Z-His-Phe-Trp-OMe	0.15	0.20	2.25

The following compounds are resistant to cleavage by pepsin:
16. Z-His-Val-Phe-OMe
17. Z-His-Ileu-Phe-OMe
18. Z-His-Ala-Ala-OMe

[a] All kinetics at 37° from references 64, 76, 79, and 80. Abbreviations used are listed in *Biochemistry*, **5**, 2485 (1966). Cha is β-cyclohexyl-L-alanyl.

TABLE 3 KINETICS OF THE PEPSIN CATALYZED HYDROLYSIS OF CATIONIC SUBSTRATES AT pH 4.0 AND 37°[a]

Substrate	k_{cat} (sec^{-1})	K_m (mM)	k_{cat}/K_m (sec^{-1} mM^{-1})
1. Z-HisPhe(NO$_2$)$\nwarrow$Phe-OMe	0.29	0.46	0.63
2. Z-His-Phe(NO$_2$)$\nwarrow$Phe-Ala	3.0	0.75	4.0
3. Z-His-Phe(NO$_2$)$\nwarrow$Phe-Ala-OMe	3.3	0.40	8.2
4. Z-His-Phe(NO$_2$)$\nwarrow$Phe-Ala-Ala-OMe	28.0	0.12	233
5. Phe-His-Phe(NO$_2$)$\nwarrow$Phe-OMe	0.008	0.17	0.04
6. Phe-Gly-His-Phe(NO$_2$)$\nwarrow$Phe-OMe	0.10	0.40	0.25
7. Phe-Gly-Gly-His-Phe(NO$_2$)$\nwarrow$Phe-OMe	0.27	0.56	0.47
8. Phe-Gly-His-Phe(NO$_2$)$\nwarrow$Phe-Ala	1.65	1.20	1.4
9. Phe-Gly-His-Phe(NO$_2$)$\nwarrow$Phe-Ala-Ala	19.2	0.27	71
10. Phe-Gly-His-Phe(NO$_2$)$\nwarrow$Phe-Ala-Ala-OMe	27.8	0.16	174

[a] Reference 82; Phe(NO$_2$) is p-nitro-L-phenylalanyl.

TABLE 4 KINETICS OF THE PEPSIN CATALYZED HYDROLYSIS OF CATIONIC SUBSTRATES[a] AT pH 2.0

Substrate	k_{cat} (sec^{-1})	K_m (mM)	k_{cat}/K_m (sec^{-1} mM^{-1})
1. Z-Phe $\prec$ Phe-OP4P[b]	0.49	0.7	0.70
2. Z-Phe $\prec$ Phe-OM3P[b]	0.14	0.81	0.17
3. Z-Gly-Phe $\prec$ Phe-OP4P	2.2	1.1	2.0
4. Z-Gly-Gly-Phe $\prec$ PheOP4P	56.5	0.8	70.6
5. Gly-Gly-Phe $\prec$ Phe-OP4P	1.5	1.4	1.1
6. Z-Ala-Ala-Phe $\prec$ Phe-OP4P[c]	260	0.04	6500

[a] Reference 83.
[b] OP4P is 3-(4-pyridyl)propyl-1-oxy and OM3P is 3-pyridylmethoxy.
[c] Reference 3, pH 3.5.

there are *large changes* in the catalytic rate constant (k_{cat}); second, the variations in the values of K_m are relatively small ($\pm$factor 10); and third, the k_{cat}/K_m figures do not indicate detectable nonproductive binding of the substrate to pepsin [81]. The only exception to this third generalization is substrate 8 in Table 2 (compare with substrate 15, Table 2), which during pepsin hydrolysis does not follow Michaelis-Menten kinetics. The explanation for this nonproductive binding in number 8, Table 2 is that Phe is preferred in the X position so that Z-His-Trp $\prec$ Phe-OMe reverses its binding to pepsin in order to place the Phe group in the X position. The complex so formed then is not in a position to be hydrolyzed by pepsin. Correlations of values of the catalytic constant k_{cat}/K_m are often made in enzyme kinetics because (1) it is easily measured for substrates of limited solubility; (2) it is independent of nonproductive enzyme substrate binding when Michaelis-Menten kinetics are followed [77]; and (3) the pH dependency of this constant has a special interpretation (*cf.* the section on pH dependency).

The data in Table 2 agree with and significantly extend the results of Table 1.

1. For substrates of the type Z-His-X $\prec$ Y-OMe aromatic amino acids are preferred in the X, Y positions, and in particular, Phe for the X position and Tyr, Phe, or Trp in the Y position. When neither position is occupied by a aromatic amino acid, the substrate is catalytically inert. However, based on the type of cleavage observed in proteins [71] Z-His-Leu-Leu-OMe may be expected to be hydrolyzed by pepsin.

2. Planar aromatic groups are preferred as side chains on either side of the cleaved bond. The substitution of the β-cyclohexyl group (number 6 and

13, Table 2) at either the X or Y position significantly diminishes the rate constant.

3. The substitution of a straight side chain amino acid in either the X or Y position (*cf.* numbers 1, 2, 3, 4, 9, 10, 11, and 12 of Table 2) yields a substrate that is still susceptible to hydrolysis. However, the substitution of a branched chain into the X position (*cf.* numbers 16 and 17, Table 2) produces an inert compound. The similar substitution into the Y position yields a substrate with diminished activity. This must mean that the side chain of the amino acid, especially in the X position, can become involved in steric hinderance.

4. The specificity requirements of the X position in compounds of the type Z-His-X ⤙ Y-OMe are most important for determining whether the substrate will be hydrolyzed by pepsin.

Tables 3 and 4 contain selected kinetic data that have been used to evaluate the influence of secondary interactions on pepsin catalyzed hydrolyses for substrates of the type A-X ⤙ Y-B. The structural features of A and B that promote hydrolysis of the X ⤙ Y peptide bond are summarized as follows:

1. As the hydrophobic character of the A and B groups increases, the compounds become progressively better substrates for pepsin. Again, this is mainly reflected in the k_{cat} value, with only small changes in K_m.

2. Compounds with a protonated terminal amino acid in close proximity to the susceptible bond are poor substrates. As this residue is moved further from the susceptible bond by the insertion of Gly residues, the substrates (*cf.* Table 3, numbers 5, 6, and 7) become progressively better. Clearly, this sensitivity of pepsin toward cleaving peptide bonds in close proximity to a protonated amino group could account for the surprising inertness of certain peptide bonds in proteins [71].

3. By systematically studying the substrates listed in Tables 3 and 4, Fruton and co-workers [82] have been able to tailor-make model substrates for pepsin. For example, the most active substrate for pepsin is Z-Ala-Ala-Phe ⤙ Phe-OP4P (*cf.* number 6, Table 4). It has also been suggested that secondary interactions of a polypeptide with pepsin play an important role in positioning the susceptible bond at the active site. This could result in a bond being either cleaved by or resistant to pepsin action. A specific example is Z-His-Phe(NO_2) ⤙ Gly-Ala-Ala-OMe, which is readily hydrolyzed, whereas Z-His-Phe(NO_2)-Gly-OMe is stable to pepsin [82].

At present it is not clear by which precise mechanism these secondary interactions increase the catalytic efficiency of pepsin. Most any of the current theories used to explain enzyme efficiency could be applied to pepsin. These could include orientation effects, rack and/or strain effects, solvent effects, and microscopic environment effects [84, 85].

3.1 Kinetic Inhibition Studies

Tables 5, 6, and 7 summarize the inhibition studies carried out on pepsin. The types of inhibition observed for pepsin are listed below [92, 93] (eq. 1–5) along with the proposed pathway(s) describing pepsin action:

Equation 1 is identical to that proposed for α-chymotrypsin, which involves the formation of one enzyme substrate covalently bound intermediate. For pepsin there is extensive evidence that this covalent bond is an amino-pepsin linkage (pepsin-NHY); that is, the acyl part of the peptide bond (X) is liberated, whereas the amino portion (Y) becomes covalently attached. Equation 2 is just an extension of equation 1 involving two pepsin-substrate covalently bound intermediates. In this proposed scheme both the acyl and the amino moieties of the peptide bond become initially attached to pepsin.

$$E + X {\small\curlywedge} Y \rightleftharpoons E{\cdot}X {\small\curlywedge} Y \rightleftharpoons E{-}Y \rightleftharpoons E \qquad (1)$$
$$\qquad\qquad\quad + X \qquad + Y \quad \text{No inhibition}$$

$$E + X {\small\curlywedge} Y \rightleftharpoons E{\cdot}X {\small\curlywedge} Y \rightleftharpoons E\overset{\displaystyle X}{\underset{\displaystyle Y}{\big\langle}} \rightleftharpoons$$

$$E{-}Y \rightleftharpoons E \qquad\qquad (2)$$
$$+ X \qquad + Y \quad \text{No inhibition}$$

$$
\begin{array}{l}
E + X {\small\curlywedge} Y \rightleftharpoons E{\cdot}X {\small\curlywedge} Y \rightleftharpoons EY \rightleftharpoons E \\
 \Big\updownarrow_{\!\!I} \qquad\qquad\qquad\quad + X \quad + Y \\
EI
\end{array}
\qquad
\begin{array}{l}
\text{Inhibition is linear} \\
\text{competitive—where} \\
\text{I can be either an} \\
\text{added inhibitor or} \\
\text{(Y)}
\end{array}
\qquad (3)
$$

$$E{-}Y{\cdot}I$$
$$\Big\updownarrow$$

$$
\begin{array}{l}
E + X {\small\curlywedge} Y \rightleftharpoons E{\cdot}X {\small\curlywedge} Y \rightleftharpoons E{-}Y \rightleftharpoons E \\
 \Big\updownarrow_{\!\!I} \qquad\qquad\qquad\quad + X \qquad + Y \\
EI
\end{array}
\qquad
\begin{array}{l}
\text{Inhibition is linear} \\
\text{noncompetitive—} \\
\text{where I can be} \\
\text{either an added} \\
\text{inhibitor or (X)}
\end{array}
\qquad (4)
$$

$$E{\cdot}I \rightleftharpoons E{\cdot}I{\cdot}I$$
$$\Big\updownarrow_{I}$$
$$\qquad\qquad\qquad\qquad\qquad\qquad\qquad \text{Mixed inhibition}$$

$$E + X {\small\curlywedge} Y \rightleftharpoons E{\cdot}X {\small\curlywedge} Y \rightleftharpoons E + X + Y \qquad (5)$$
$$ \Big\updownarrow_{\!\!I}$$
$$E{\cdot}I{\cdot}X {\small\curlywedge} Y$$

TABLE 5 COMPETITIVE INHIBITION CONSTANTS OF SUBSTRATE LIKE INHIBITORS (COMPARISON WITH THE APPROPRIATE MICHAELIS CONSTANTS)

Inhibitor[a]	K_I (mM)[b]	K_m (mM)[c]	pH	Temperature	Reference
1. Ac-*Phe*-*Tyr*-OMe	2.3	1.5	4.05	25°	86
	1.9	2.4	2.10	25°	86
	2.5	1.3	1.05	25°	86
2. Z-His-*Phe*-Phe-OEt	0.175[d]	0.18	4.0	37°	64
	0.20[e]				
3. Z-His-Phe-*Phe*-OEt	0.275[d]	0.18	4.0	37°	64
	0.275[e]				
4. Ac-*Phe*-Phe	1.8 (1.5)	1.5	2.2	37°	87
	6.0 (6.2)		4.3	37°	87
5. AcPhe-*Phe*	1.3 (1.5)	1.5	2.1	37°	87
	5.7 (5.5)		4.3	37°	87
6. Ac-*Phe*-*Phe*	0.76	1.5	2.2	37°	87
7. Ac*Phe*-PheNH$_2$	0.61 (1.5)		2.0	37°	87
	1.7 (3.0)		4.7	37°	87
8. Ac-Phe-*Phe*NH$_2$	0.5 (0.5)		2.0	37°	87
	2.2 (2.2)		4.7	37°	87
9. Ac-*Phe*-Pla	1.3	1.4	2.2	37°	87
10. Z-His-Phe(NO$_2$)-Pol[f]	0.1		4.0	20°	88
11. Ac-AmCinn-diodoTyr[g]	≈0.09 (0.06)	0.075[h]	2.0	35°	89
12. Ac-AmCinn-Tyr[i]	≈(3)	1.2	2.0	35°	89

[a] The amino acid in the D-configuration is in italics.

[b] The value of the inhibition constant in parentheses was obtained by a nonkinetic method (i.e., equilibrium dialysis or gel filtration).

[c] When available the kinetically determined binding constant for the catalytically reactive L-L substrate is shown for comparison.

[d] Substrate Z-His-Phe(NO$_2$) ⇌ PheOMe.

[e] For comparison the substrate was the depsipeptide Z-His-Phe(NO$_2$)-Pha-OMe: Pha is L-β-phenyllactyl.

[f] Pol is L-phenylalaninol. A comparison substrate, Z-His-Phe(NO$_2$)-Phe-OMe, exhibits a K_m of 0.46 mM at 37°. The inhibition constant was determined by the gel filtration method.

[g] Substrate is N(α-Acetamidocinnamoyl)-L diiodotyrosine. Solvent 3.4%. methanol. The K_I in parenthesis was determined spectrophotometrically. This compound behaves as a noncompetitive inhibitor.

[h] The substrate for comparison is Ac-Phe-diiodo-Tyr 37°, no organic solvent present.

[i] The substrate for comparison is Ac-Phe-Tyr. Solvent 3.4% methanol.

TABLE 6 INHIBITION CONSTANTS FOR COMPOUNDS WHICH ARE TYPICALLY PRODUCTS OR PRODUCT ANALOGUES OF PEPSIN HYDROLYSES

Inhibitor	Type of Inhibition	K_I (mM)	pH	Temperature	Reference
1. Ac-Phe	l-noncompetitive[d]	28 (28)[b]	2.1	37°	91, 92
2. Ac-Phe	l-noncompetitive	23	2.0	37°	72, 93
3. Ac-Phe	l-competitive	55	4.0	37°	97
4. Ac-Phe	l-competitive	100[a]	4.0	25°	23
5. Ac-Tyr	Mixed	41	2.0	37°	93
6. Ac-Tyr	l-competitive	62	4.0	37°	97
7. Ac-Ala	l-competitive	151	2.0	37°	93
8. Ac-Gly	competitive	366	2.0	37°	93
9. Ac-Trp	l-competitive	35	4.0	37°	97
10. Ac-Leu	l-competitive	70	4.0	37°	97
11. Ac-Phe-OEt	l-noncompetitive	17 (12)[b]	2.0	37°	92
12. Ac-Phe-NH₂	l-noncompetitive	47 (46)[b]	2.0	37°	92
13. MeSul-Phe-OMe	l-noncompetitive	15	2.0	37°	91
14. MeSul-Phe-SMe[e]	l-noncompetitive	98	2.0	37°	91
15. Phe	l-competitive	90	4.0	37°	97
16. Phe-OL[c]	l-competitive	5	4.0	37°	97
17. Phe-OMe	l-competitive	22	4.0	37°	97
18. Phe-OMe[a]	l-competitive	8	2.1	25°	23
19. Phe-OEt	l-competitive	10 (10)[b]	4.0	37°	97
20. Leu-OEt	l-competitive	19	4.0	37°	97
21. Tyr-OEt	l-competitive	18	4.0	37°	97
22. Tyr-OMe[a]	l-competitive	23	1.05	25°	23
23. Trp-OEt	l-competitive	6	4.0	37°	97
24. His-Phe	l-competitive	17	4.0	37°	97
25. Phe-Phe	l-competitive	1	4.0	37°	97
26. Phe-Phe-OMe	l-competitive	0.25	4.0	37°	97
27. Phe-Tyr-OMe	l-competitive	0.86	4.0	37°	97
28. Z-His-Phe-Phe	l-competitive	0.75	4.0	37°	97

[a] 3% dioxane.
[b] Number in parenthesis is for the D amino acid. Unlike the L isomer, these D compounds act as linear competitive inhibitors.
[c] L-Phenylalaninol.
[d] Linear noncompetitive.
[e] MeSul represents methanesulfonyl.

TABLE 7 INHIBITION CONSTANTS OF SELECTED PEPSIN INHIBITORS

Inhibitor	Type of Inhibition	K_I (mM)	pH	Temperature	Reference
1. MeOH	Competitive	620	3.5	35°	96
2. EtOH	Competitive	214	3.5	35°	96
3. n-PrOH	Competitive	92	3.5	35°	96
4. n-BuOH	Competitive	33	3.5	35°	96
5. i-BuOH	Competitive	26	3.5	35°	96
6. t-BuOH	Competitive	45	3.5	35°	96
7. n-Amyl-OH	Competitive	16	3.5	35°	96
8. Dioxane	Competitive	622	2.1	25°	23
9. Phenol	Mixed	23.4	2.0	37°	93
10. 2,4,6-TriMe-phenol	Mixed	5.1	2.0	37°	93
11. 4-Et-phenol	Mixed	8.5	2.0	37°	93
12. L-β-phenyllactic acid	Mixed	18.1	2.0	37°	93
13. Trans-cinnamic acid	Mixed	2.8	2.0	37°	93
14. Cyclohexanol	Mixed	22.2	2.0	37°	93
15. Benzoic acid	Mixed	12.2	2.0	37°	93

The evidence supporting the expansion of equation 1 to equation 2 to include both an acyl and amino-pepsin covalently bound intermediate is on a more tenuous foundation.

The data in Table 5 can be analyzed in the following way:

1. As has been known for some time [1] both amino acids that form the susceptible peptide bond must be of the L-configuration.

2. If either or both are D the compound is a competitive inhibitor.

3. For substrates of the type A-X ⤳ Y-B there is no information as to how D-amino acids in the A and B portions will affect the compound's reactivity with pepsin.

4. The binding constants (either K_I or K_m) for neutral compounds are essentially independent of pH in the pH region of pepsin activity.

5. For inhibitors with an ionizable carboxyl group (i.e., Ac-*Phe-Phe*) the K_I increases significantly as the pH increases. This implies that substrate-type compounds with an ionized carboxyl group bind very poorly or not at all to the active site of pepsin.

6. Finally a comparison of the K_I for the D-D, L-D or D-L inhibitor with the K_m for the corresponding L-L substrate shows them to be equal within experimental error. This equivalence has been interpreted to mean that the observed K_m is a true binding constant not modified by the various rate constants. Relating these observations to equation 1 (or eq. 2) implies that the catalytic step which immediately follows the formation of the Michaelis complex is rate limiting. The release of the cleavage products occurs after the slow step of the reaction (*cf.* section on D_2O effects).

7. Equation 3 describes the linear competitive inhibition exhibited by these compounds by binding only to the free pepsin.

An analysis of Table 6 suggests the following:

1. For compounds that are typical products or product analogs, the incorporation of D-amino acids results in their being linear competitive inhibitors (Table 6, numbers 1, 11, 12, and 19) as described by equation 3. Presumably due to the conformational restrictions imposed by the pepsin-substrate intermediates, these simple D compounds can bind only to the free enzyme. In fact, binding to the free enzyme must be the most important contribution to the magnitude of the K_I values for the corresponding L-inhibitors. This must be so since the K_I's for L and D isomers are identical, but exhibit different types of inhibition.

2. The amino terminal product or product analogs act only as linear competitive inhibitors in the pH range of pepsin action (Table 6, numbers 15–23).

3. Substratelike compounds (Table 6, numbers 24–28) with either an unprotected amino and/or carboxyl terminal group act as competitive inhibitors at pH 4.0. This type of specificity has been invoked to explain the anomalous or unexpected cleavage products from the pepsin hydrolyses of proteins [71].

4. Mechanistically, the most interesting analysis of the data from Table 6 has been reported by Knowles and co-workers [91, 92]. These workers applied to pepsin the approach exemplified by Hsu, Cleland, and Anderson [98] on the inhibition pattern of acid potato phosphomonoesterase. This analysis supports a mechanism for the *ordered release* of products from pepsin catalyzed reactions (eqs. 1, 3, and 4). These inhibition studies give additional support for the postulate of at least one pepsin-substrate covalent intermediate. At pH 2, the acyl portion of the substrate (i.e., acetyl L-phenylalanine) shows linear noncompetitive inhibition [91, 92], whereas the second product liberated, Y—the amino portion of the substrate—shows linear competitive inhibition [97]. The corresponding kinetic equations [92] are consistent with equation 4. This pathway suggests then that the first product liberated during hydrolysis, X (i.e., acetylphenylalanine) can bind to the free enzyme and unproductively to the amino enzyme (E-Y). Knowles proposes that the protonated acetylphenylalanine (X) at pH 2.0 binds nonproductively and only the corresponding anion (X⁻) at pH > 4 can bind productively. This explanation is consistent with his mechanism for pepsin action (*cf.* section on Mechanism), which demands that the X moiety of the substrate be *released as the anion*.

The above analysis can also be used to explain why the neutral compounds (Table 6, numbers 11–14) Ac-Phe-OEt and Ac-Phe-NH$_2$ act as noncompetitive inhibitors. Nonproductive binding by these neutral acyl-amino acids also would explain why they do not act as acceptors in the transpeptidation reaction (*cf.* section on transpeptidation and ^{18}O exchange). Furthermore, the acyl amino acids (Table 6, numbers 3, 4, 6, and 9), which act as linear noncompetitive inhibitors at pH 2 (where they are fully proton-ated), act as linear competitive inhibitors at pH 4.0. At pH 4.0, assuming the pK_a of the carboxyl group of an acylated amino acid to be about 3.5, only about 20 % of the inhibitor is still in the protonated form. As the pH increases, K_I also increases, suggesting that the anion form of the acyl amino acid (i.e., X^-) does not bind as well to the free enzyme. These results are especially interesting in relation to the transpeptidation reaction, which is much more important at the higher pH (cf. section on transpeptidation). It can be shown that equation 6 is consistent with the inhibitor results at pH 4.0 [92].

In fact, equation 6 is only a slightly modified form of equation 4.

$$E + X\text{-}Y \rightleftharpoons E \cdot X\text{-}Y \rightleftharpoons E\text{-}Y \rightleftharpoons E$$
$$X \big\updownarrow \qquad\qquad\qquad +X \quad +Y \qquad (6)$$
$$EX$$

Finally, it should be noted that the simple aliphatic acyl amino acids (Table 6, numbers 7 and 8) act as competitive inhibitors even at pH 2. Presumably, they bind rather poorly only to the free enzyme.

Table 7 contains a summary of the inhibition studies on pepsin, employing small molecules. The results are summarized below, and a more extensive list is found in reference [93].

1. At high concentrations of inhibitor, pepsin binds more than one molecule per molecule of enzyme. Using the gel filtration method, Hum-phreys and Fruton [88] present data suggesting that the substratelike molecule Z-His-Phe(NO_2)-Pol (cf. Table 5, number 10) binds to both the primary and secondary binding sites on pepsin.

2. Equation 5 essentially describes this type of inhibition as being com-petitive at lower inhibitor concentrations and "mixed" at higher concen-trations.

3. Table 7 also shows that alcohols competitively inhibit pepsin activity. This inhibitory activity increases with increased chain length of the hydro-carbon group of the alcohol. These results support the hypothesis that hydrophobic bonding between pepsin and the side chains of the amino acids in the substrate is the major factor in pepsin-substrate bonding [95, 96].

4. Knowles [99] has suggested that the solubility of a compound in water can be used as a reasonably good index of the compound's inherent hydro-phobic character. Schlamowitz, Shaw, and Jackson [93] have applied this

criterion to pepsin and observed a roughly linear relation between water solubility (hydrophobicity) and binding energy for 20 inhibitor compounds.

3.2 Inhibition by Chemical Modification

The chemical modification of amino acid residues provides essential information for the correlation of protein (enzyme) structure and function. In addition, these modified proteins often can be used as crystalline derivatives in the x-ray studies leading to the three-dimensional structures of the proteins.

3.2.1 Carboxyl Group. Northrop [100], as early as 1922, found that pepsin was active on protein substrates in a pH range from below 2 to above 4 with a maximum at pH 1.8. This is undoubtedly the first report that could be interpreted as implicating carboxyls as the catalytically active groups on pepsin. The first direct chemical evidence supporting this idea came from Herriott [101] who observed that the esterification of the carboxyl groups on pepsin by mustard gas ($ClCH_2CH_2SCH_2CH_2Cl$) produced an inactive enzyme. Pepsin was 63% inactivated with 5 carboxyl groups esterified. It has since become clear that side chain carboxyl groups of aspartic and/or glutamic acids in pepsin are the functional groups which directly participate in the enzymatic action.

Recent interest in using chemical reagents to study the catalytic role of the carboxyl groups in pepsin was stimulated by the work of Delpierre and Fruton [102]. They found that a fourteen fold excess of diphenyldiazomethane esterifies about two–three carboxyl groups on pepsin (at 30° pH 5.1), causing a 50% loss in activity. Because of the high chemical reactivity of diphenyldiazomethane, more selective diazo reagents were sought. The hope that these more selective diazo compounds would react with pepsin in a 1:1 molar ratio has been realized with a large number of inhibitors (Table 8). Table 8 is a listing of the 20 carboxyl group inhibitors that have been used on pepsin. The reagents were chosen because of their known reactivity with carboxyl groups to form esters. For reference, Table 9 contains typical experimental results from two studies [103, 104]. Other studies on carboxyl inhibitors used similar experimental conditions. However, there is sufficient variation in experimental details to make this tabular listing too complex. Therefore the results of all the investigations are summarized below.

1. In every inhibition study the concentration of inhibitor and other reagents was several times that of pepsin. It is interesting to note that Hamilton and co-workers [104] found that a large excess of the diazoketone

TABLE 8 CHEMICAL INHIBITORS OF PEPSIN

Compound	Name (Abbreviation)	Reference
1. 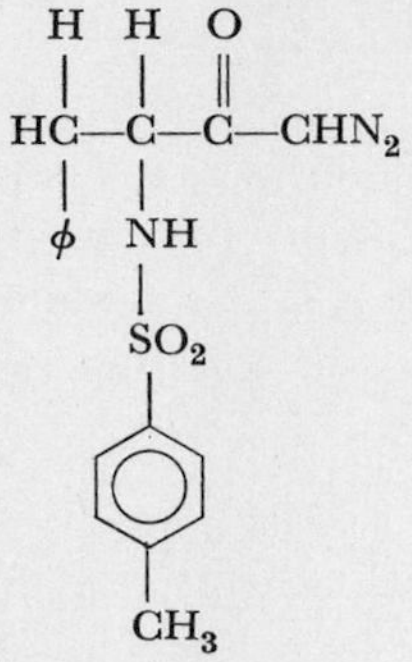 $\begin{array}{c} \text{H} \quad \text{H} \quad \text{O} \\ | \quad | \quad || \\ \text{HC—C—C—CHN}_2 \\ | \quad | \\ \phi \quad \text{NH} \\ \quad\quad | \\ \quad\quad \text{SO}_2 \\ \quad\quad | \\ \quad\quad \bigcirc \\ \quad\quad | \\ \quad\quad \text{CH}_3 \end{array}$	L-1-Diazo-4-phenyl-3-tosylamidobutanone DDTB	103
2. 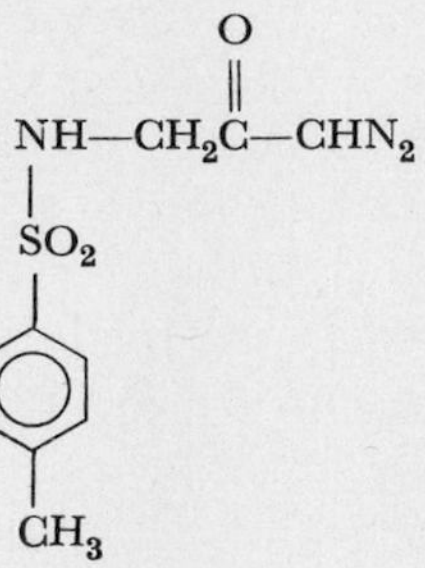 $\begin{array}{c} \quad\quad\quad \text{O} \\ \quad\quad\quad || \\ \text{NH—CH}_2\text{C—CHN}_2 \\ | \\ \text{SO}_2 \\ | \\ \bigcirc \\ | \\ \text{CH}_3 \end{array}$	1-Diazo-3-tosylamidopropanone	103
3. 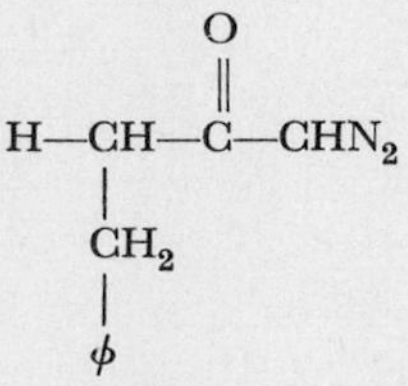 $\begin{array}{c} \quad\quad \text{O} \\ \quad\quad || \\ \text{H—CH—C—CHN}_2 \\ \quad | \\ \quad \text{CH}_2 \\ \quad | \\ \quad \phi \end{array}$	1-Diazo-4-phenylbutanone (DPB)	47, 104, 105
4. 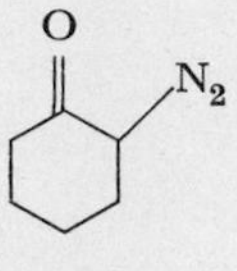 2-Diazocyclohexanone structure (N_2)	2-Diazocyclohexanone (DCH)	104
5. 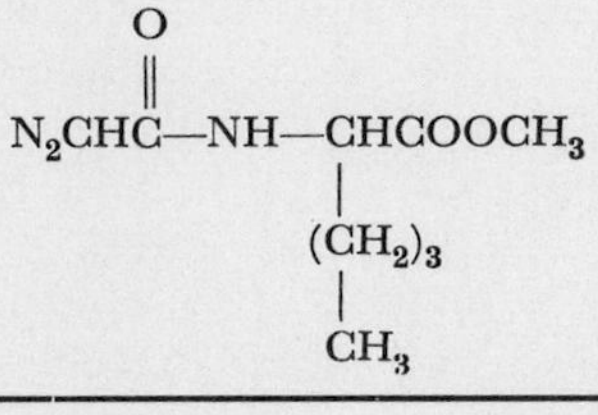 $\begin{array}{c} \text{O} \\ || \\ \text{N}_2\text{CHC—NH—CHCOOCH}_3 \\ \quad\quad\quad\quad | \\ \quad\quad\quad\quad (\text{CH}_2)_3 \\ \quad\quad\quad\quad | \\ \quad\quad\quad\quad \text{CH}_3 \end{array}$	Diazoacetylnorleucine methyl ester	107

Table 8 (*cont.*)

Compound	Name (Abbreviation)	References

6. $N_2CHC(=O)—NH(CH_2)_2$ — (2,4-dinitrophenyl ring with NO_2) — *N*-diazoacetyl N'-2,4 dinitrophenylethylene diamine (DDE) — 18, 106

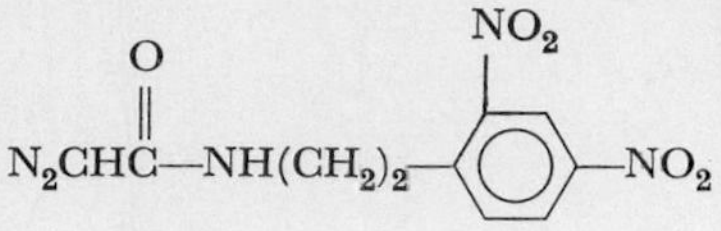

7. $N_2CHC(=O)NHCH_2COOC_2H_5$ — Diazoacetyl glycine ethyl ester (IGG) — 109, 113

8. $N_2CHC(=O)—OC_2H_5$ — Diazoacetic acid ethyl ester — 109

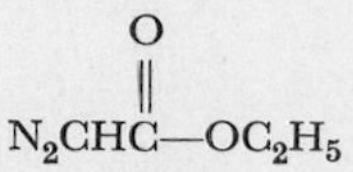

9. $\phi—C(O^-)=C(H)—\overset{\oplus}{S}(CH_3)_2$ — Dimethylsulfonium phenacylide — 109

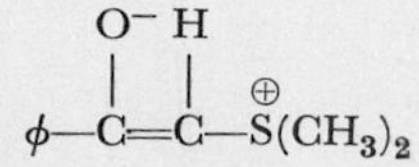

10. $\phi—CH_2—CH(NHC(=O)—OCH_2\phi)—C(=O)CHN_2$ — Benzyloxycarbonyl-L-phenylalanyldiazomethane (ZPDM) — 110

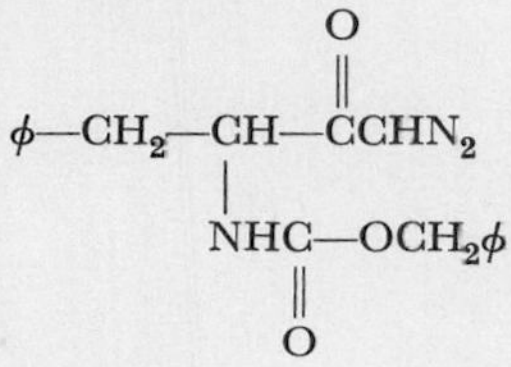

11. $N_2CHC(=O)—NH—CH(CH_2\phi)—COOCH_3$ — *N*-Diazoacetyl-L-phenylalanine methyl ester (IGP) — 111, 112 / 113

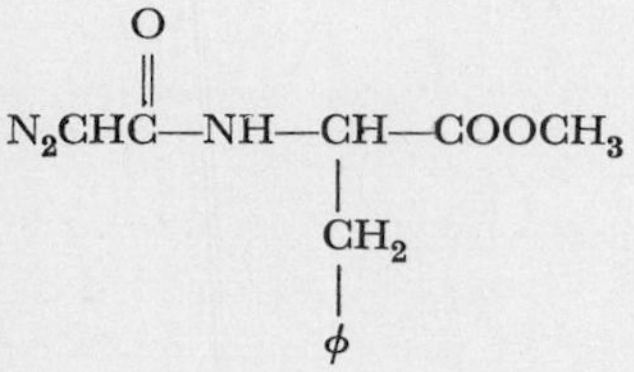

12. $N_2CHC(=O)—CH_2NH$ — (2,4-dinitrophenyl ring with NO_2) — 1-Diazo-3-dinitrophenyl-aminopropanone-2 (IKG)

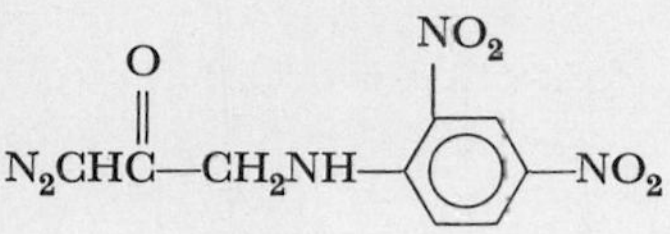

13. $C_2H_5O_2CC(N_2)—CH_2$ — (phenyl ring) — OH — α-Diazo-β-p-hydroxyphenyl propionic ethyl ester — 113

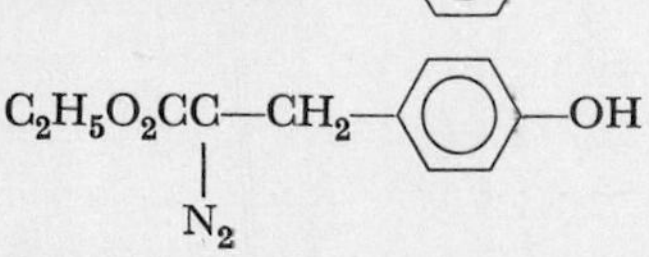

14. $\phi C(=O)—C(N_2)—\phi$ — Phenylbenzoyl diazomethane — 113

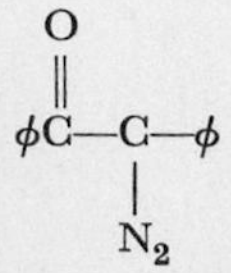

Table 8 (cont.)

Compound	Name (Abbreviation)	References
15. 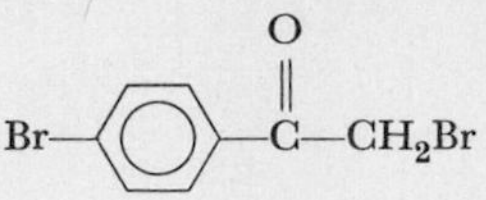 Br—⟨○⟩—C(=O)—CH$_2$Br	p-Bromophenacylbromide	114–117
16. 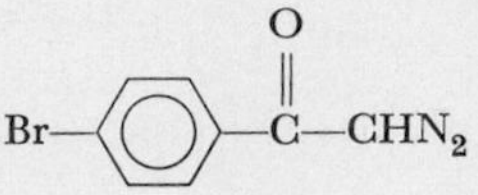 Br—⟨○⟩—C(=O)—CHN$_2$	α-Diazo-p-bromoacetophenone	117
17. φ(CH$_2$)$_2$C(=O)—CHN$_2$	1-Diazoacetyl-2-phenylethane	118
18. φCH$_2$CH(C(=O)—CHN$_2$)$_2$	1,1-Bis-(diazoacetyl)2-phenylethane	118
19. φCH$_2$CHC(=O)—CHN$_2$ (with Br on middle C)	dl-1-Diazoacetyl-1-bromo-2-phenylethane	118
20. 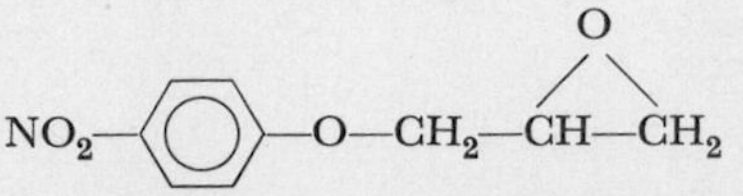 NO$_2$—⟨○⟩—O—CH$_2$—CH—CH$_2$ (epoxide)	1,2-Epoxy-3-(p-nitrophenoxy) propane (EPNP)	119

inhibitor number 3 (Table 8) actually decreased the rate of inhibition of pepsin (Table 9, Part B). An excess of inhibitor seems to be required because of its significant side reaction with water. However, these diazo compounds should make excellent reagents for a titration of the operational normality of a pepsin solution. Certainly, with the vast quantity of data acquired on pepsin modification, the exact experimental conditions for a pepsin titration should soon be forthcoming.

2. With the diazo compounds, Cu^{2+} markedly catalyzes the inhibition of pepsin. In the absence of Cu(II) there is slow inactivation of pepsin, but the rate is generally not very reproducible. The pH-inactivation profile is a sigmoid curve in the presence of Cu(II) with a maximum inactivation rate achieved above pH 5.5 [109, 113]. The prior mixing of Cu(II) and the diazo inhibitor significantly increases (about a factor of 2) the rate of pepsin inactivation [109, 113]. In particular, prior incubation of Cu(II) with IGG (Table 8, number 7) not only increases the rate of pepsin inactivation but changes the pH-inactivation curve to bell-shaped and lowers the pH

TABLE 9   INACTIVATION OF PEPSIN BY DIAZOKETONES

Part A

Reactants[a,b]		Fraction Pepsin Activity Lost[c]		Equivalent Incorporation of Inhibitor[h]	
		5 min	45 min	5 min	45 min
L-DPTB	Pepsin	0.51	0.94	0.47	0.99
	Pepsinogen	0	0	0.01	0.02
	Denatured pepsin	—	—	0.03	0.10
D-DPTB	Pepsin	0	0.61	—	—
Tos-Gly-CHN$_2$	Pepsin	0.11	0.82	0.12	0.82

Part B[i]

Concentration of Reagents $\times$ 10^4 (M)		Inhibitor (concentration $\times$ 10^4)	Reaction Time (min)	% Inhibitor[h] Incorporation	% Pepsin Inhibited[e]
Pepsin	Cu(II)				
6.5	7	DPB[d](18)	120	72	86
9.3	1.0	DPB(10)	30	81	91
9.3	1.0	DPB(100)	30	60	60
4.6[f]	0.5	DPB(5)	120	3	—
0.3	0.0	DPB(0.83)	120	—	40
0.3	0.5	DCH[g](0.66)	120	—	0

[a] Reference 103.
[b] Pepsin conc. 0.0286 mM, diazoketone. 0.143 mM; CuCl$_2$ 1 mM pH 5.3; 15°.
[c] Assayed against hemoglobin.
[d] 1-Diazo-4-phenylbutanone; pH 5.5; 38°.
[e] Assayed vs. Ac-Phe-Tyr at pH 2, 38°.
[f] Pepsinogen.
[g] 2-Diazocyclohexanone.
[h] The equivalents of inhibitor incorporated into pepsin were determined using ^{14}C labeled diazo compounds.
[i] Reference 104.

$$R-\overset{\overset{\displaystyle O}{\|}}{C}-CHN_2 + Cu(II) \longrightarrow R-\overset{\overset{\displaystyle O}{\|}}{C}-CH=Cu(II) + N_2 \qquad (7)$$

maximum to 5 [109]. It has been proposed [108, 109, 121] that Cu(II) reacts with the diazo reagents to form a metal-complexed carbene intermediate according to equation 7. Presumably, the prior mixing allows the reaction shown in equation 7 to be complete before the pepsin is added. The positively charged copper carbene intermediate then binds to a negatively charged carboxyl group at or near the active site. In a second step this complexed pepsin-inhibitor intermediate reacts to esterify a neighboring protonated carboxyl group, which is presumably in the active site (equation 8). Although the optimal experimental conditions have not been determined, dimethylsulfonium phenacylide (number 9, Table 8), which requires Cu(II), also reacts to inactive pepsin [109, 121].

3. The metal ions, Cd(II), Co(II), Pb(II), Zn(II) [109] and Ni(II) [113], are not effective in promoting this inactivation reaction. There is one report [109] that Ag(I) is as effective as Cu(II) with diazoacetylnorleucine methyl ester; another report states that Ag(I) with diazoacetylglycine ethyl ester does not inhibit pepsin [113].

4. There are two major pieces of evidence that the inhibition of pepsin by diazo compounds is a specific enzymatic reaction. First, little or no reaction occurs with base-denatured pepsin or native pepsinogen [103, 107]

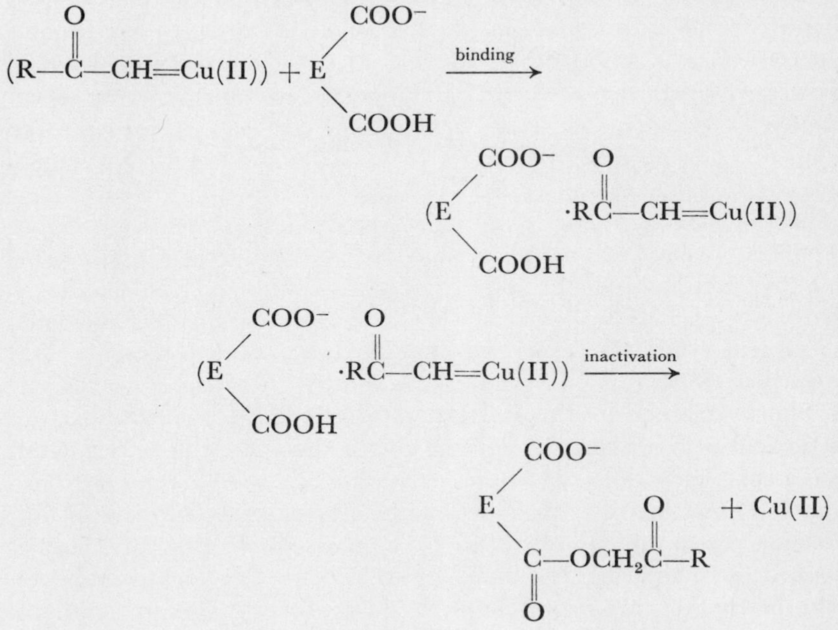

$$(R-\overset{\overset{\displaystyle O}{\|}}{C}-CH=Cu(II)) + E\overset{\displaystyle \diagup COO^-}{\underset{\displaystyle \diagdown COOH}{}} \xrightarrow{\text{binding}}$$

$$(E\overset{\displaystyle \diagup COO^-}{\underset{\displaystyle \diagdown COOH}{}} \quad \cdot R\overset{\overset{\displaystyle O}{\|}}{C}-CH=Cu(II))$$

$$(E\overset{\displaystyle \diagup COO^-}{\underset{\displaystyle \diagdown COOH}{}} \quad \cdot R\overset{\overset{\displaystyle O}{\|}}{C}-CH=Cu(II)) \xrightarrow{\text{inactivation}} \qquad (8)$$

$$E\overset{\displaystyle \diagup COO^-}{\underset{\displaystyle \diagdown C-OCH_2\overset{\overset{\displaystyle O}{\|}}{C}-R}{}} \overset{\displaystyle}{} + Cu(II)$$

(cf. Table 9), nor is the potential catalytic activity of pepsinogen affected. The labeled group is readily removed from pepsin in the pH range 7–8 by reaction with water [47, 112] or the added nucleophile, hydroxylamine [107, 113]. This observed lability of the enzyme-inhibitor bond suggests an ester linkage. Second, evidence that binding at a specific site on the enzyme is necessary for inactivation is based on the observation that 2-diazocyclohexanone (cf. number 4, Table 8 and Table 9, Part B) does not inhibit pepsin; D-DPTB inhibits pepsin more slowly than the L-isomer (Table 9, Part A) and the nonspecific diazoacetic acid ethyl ester (number 8, Table 8) reacts to incorporate more than one equivalent of inhibitor.

5. Unless otherwise indicated, the diazo compounds listed in Table 8 react with pepsin in a 1:1 molar ratio. The parallelism between equivalents of inhibitor incorporated and loss of proteolytic activity is readily seen in Table 9. Parallelism of activity was maintained whether the rate assay was carried out with hemoglobin, Z-His-Phe-Try-OEt (a very reactive substrate), or Ac-Phe-Tyr-NH$_2$.

6. The use of these stoichiometric active-site-directed irreversible diazo inhibitors has made possible sequence studies around a reactive aspartic acid residue. Three separate research groups using three different inhibitors give the following sequences: Val-**Asp** [106], Ile-Val-**Asp**-Thr-Gly-Thr-Ser [112], and Ile-Val-**Asp**-Thr-(Gly, Thr)-Ser-Leu [47]. There seems little doubt that when one carboxyl group of a specific aspartic acid in the foregoing sequence of pepsin is esterified by a diazo inhibitor, the enzyme becomes completely inactive. For further details concerning this reactive aspartic acid residue, refer to the section on pH dependency of pepsin action.

7. At the same time that the diazo inhibitors of pepsin were being tested, other types of compounds capable of esterifying carboxyl groups were also developed. Erlanger and co-workers [114, 115] observed that p-bromophenacylbromide is a specific inhibitor of pepsin reacting in a mole/mole ratio. The inactivation process can be competitively inhibited by Z-Phe or Ac-Phe-Tyr. The pH rate profile for inactivation is pseudo-bell-shaped [114] with a maximum inhibition in the range pH 2–3. Although the inactivated pepsin was completely inert to Z-Glu-Phe, its inactivation toward hemoglobin never exceeded 78 %. Gross and Morrell [116] showed that p-bromophenacyl bromide inactivated pepsin by esterifying a carboxyl group of an aspartic acid. Shortly thereafter Erlanger and co-workers [115] reported that this aspartic acid is in a sequence with the composition (Gly$_2$, Asp, Ser, Glu). This is a completely different sequence than that observed around the diazo reagent inhibited aspartic acid. This apparent discrepancy was resolved [117] by treating pepsin with p-bromophenacyl bromide followed by diazo bromoacetophenone (numbers 15 and 16, Table 8). The diazo compound completely inactivates pepsin and two p-bromophenacyl moieties become

attached to pepsin. These results imply that the aspartic acid inactivated by phenacyl bromide does not participate directly in the catalytic mechanism but only interferes stereochemically with the enzyme-substrate interaction.

8. Gross [122] has used this specific esterification reaction of pepsin to achieve a nonenzymatic fragmentation of a single peptide bond in this enzyme. This fragmentation reaction is shown in the following equations. Because two different aspartic acid side chains have been separately esterified, this reaction would allow for two points of cleavage. Surely, this sort of reaction could be applied to sequence work on other proteins (enzymes).

9. Tang [119] has used still another esterifying reagent, an epoxide

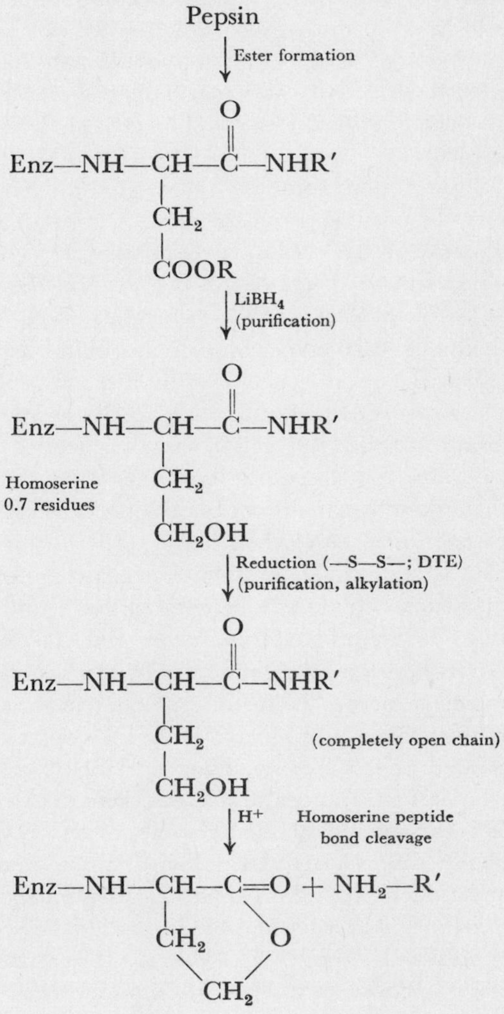

$$
\begin{array}{c}
\text{Pepsin} \\
\downarrow \text{Ester formation} \\
\\
\text{O} \\
\parallel \\
\text{Enz—NH—CH—C—NHR}' \\
| \\
\text{CH}_2 \\
| \\
\text{COOR} \\
\downarrow \text{LiBH}_4 \\
\text{(purification)} \\
\\
\text{O} \\
\parallel \\
\text{Enz—NH—CH—C—NHR}' \\
| \\
\text{Homoserine} \quad \text{CH}_2 \\
\text{0.7 residues} \quad | \\
\text{CH}_2\text{OH} \\
\downarrow \text{Reduction (—S—S—; DTE)} \\
\text{(purification, alkylation)} \\
\\
\text{O} \\
\parallel \\
\text{Enz—NH—CH—C—NHR}' \\
| \\
\text{CH}_2 \quad \text{(completely open chain)} \\
| \\
\text{CH}_2\text{OH} \\
\downarrow \text{H}^+ \quad \text{Homoserine peptide} \\
\text{bond cleavage} \\
\\
\text{Enz—NH—CH—C=O} + \text{NH}_2\text{—R}' \\
| \qquad \quad \backslash \\
\text{CH}_2 \qquad \text{O} \\
\backslash \qquad / \\
\text{CH}_2
\end{array}
$$

(number 20, Table 8). Epoxides are known to react with carboxylates in acid solution [123]. When pepsin is treated with this epoxide, two moles are incorporated per mole of inactivated enzyme. As with previous inhibited pepsins, these covalently bound residues are removed in alkaline solution by base catalyzed ester hydrolysis. Prior reaction of pepsin with diazoacetyl-DL-norleucine methyl ester or p-bromoacetophenone still resulted in the incorporation of two epoxide residues per molecule of pepsin. Therefore, the two carboxyls inhibited by the epoxide must be different from those previously mentioned. The addition of substrate protected one of the carboxyls from reaction with the epoxide reagent. Possibly this protected carboxyl is part of the active site of pepsin and participates in catalysis by the enzyme. The second carboxyl group not protected by the substrate apparently is not essential for enzyme activity. Refer to the following pH-dependency section for a discussion of the requirement for two catalytically important carboxyl groups in the active site of pepsin.

10. Husain, Ferguson, and Fruton [118] have extended this inhibition reaction one step further by developing bifunctional inhibitors (numbers 17, 18, and 19, Table 8). These reagents react stoichiometrically to inhibit pepsin, and the inhibited enzyme has approximately the same molecular weight as pepsin. Evidence was presented to show that at least some of the bifunctional reagent has formed two bonds to the same pepsin molecule.

11. Finally, Hoare and Koshland [120] suggest modifying carboxyl groups in proteins by using a water-soluble carbodiimide (N-benzyl-N'-3-dimethyl-aminoproplycarbodiimide) as the activating agent and glycine methyl ester as the modifying reagent. These reagents have not been used on pepsin. However, the proper experimental conditions could probably be worked out with these reagents to gain additional information about the very important carboxyl groups in pepsin.

In summary, pepsin has one active site aspartic acid carboxyl group inhibited by diazo compounds; another active site carboxyl inhibited by an epoxide; and two reactive nonactive site carboxyls, one inhibited by the epoxide and the other inhibited by the bromoketone.

3.2.2 Tyrosine. Historically, the first studies on the modification of amino acids in pepsin dealt with the hydroxyl group of tyrosine and not carboxyl groups. The increasing substitution of the phenolic groups by acylation with ketene caused progressive inhibition of pepsin toward hemoglobin [124]. With a more selective acylating reagent, acetyl imidazole, acylation of the tyrosine groups of pepsin at pH 5.5 and 5.8 results in a loss of proteinase activity toward hemoglobin but an increased peptidase activity toward synthetic peptides [125, 60]. The restoration of proteolytic activity occurs on standing in the presence of imidazole by deacylation of 4-5 phenolic groups [125]. This increased peptidase activity is a consequence of increased catalytic efficiency with only small changes in the Michaelis constant [60]. An explanation for the loss of proteolytic activity is that the secondary binding sites on pepsin are blocked as a result of acylation, which prevents effective cleavage of long chain peptides (60).

During the reaction of pepsinogen with acetyl imidazole, more than 10 phenolic groups are acylated, but this has no effect on the activation of pepsinogen to pepsin. Similar results have been reported for the nitration of pepsin with tetranitromethane [126] and carbamylation with potassium cyanate [127]. In the carbamylation reaction six tyrosines and the N-terminal isoleucine react, but the one lysine was not carbamylated. In contrast to acylation, iodination of pepsin results in a parallel loss of proteolytic, peptidase, and esterase activity [60]. Because the iodination reaction might also effect imidazole and tryptophan groups, the interpretation of this reaction is ambiguous.

3.2.3 Tryptophan and Methionine.

Dopheide and Jones [44] treated pepsin with Koshland's tryptophan modifying reagent, 2-hydroxy-5-nitrobenzyl bromide. At pH 3.5, active pepsin incorporates two inhibitor residues leading to a 25–30% loss of proteolytic or peptidase activity. This result suggests that tryptophan is not an essential amino acid directly participating in the catalytic process. In alkali denatured pepsin three tryptophan residues are accessible, but the fourth reacted only after cleavage of the disulfide bonds. The total number of tryptophan residues in pepsin is still in doubt, with reports varing from 4 to 6.

Alternatively, N-bromosuccinimide (NBS) also inactivates pepsin, by destruction of the tryptophan and methionine residues. Apparently tyrosine is inert toward bromination by NBS [128]. Synthetic substrates and some simple amino acids (tryptophan, methionine, or tyrosine) protect pepsin from inactivation by NBS [128]. Dopheide and Jones [44] note that the methionine residues are not alkylated by iodoacetate and iodoacetamide. However, Lokshina and Orekhovich [128] find that methionine is alkylated by iodoacetate with no loss in pepsin activity. Finally, the chromophoric group (2,4-dinitrophenyl, 1-dimethylnaphthalene-5-sulfonyl-) has been introduced into pepsin to study resonance interactions with the tryptophan moieties [129].

3.2.4 Disulfide Bridges.

Pepsin contains six cysteine residues that in the native enzyme form three disulfide bridges. Pepsinogen also contains three disulfide bonds, and the amino acid sequence around these cysteines is identical in the enzyme and its precursor. Nakagawa and Perlmann [130] reported the following amino acid sequence around these disulfide bridges:

Disulfide bridge A

—Glu-Asx-Asx-Ser-Cys-Thr-Ser-Asp-Ser-Asp-Ser—
     |
    —Cys-Ser-Ser-Ile-Asp-Gln—

Disulfide bridge B

—Cys-Ser-Gly-Cys-Gln—

Disulfide bridge C

—Cys-Ser-Ser-Leu-Ala-Cys-Ser-Asp-His-Asn-Gln-Phe—

Reduction of these disulfide bonds with β-mercaptoethanol occurs in an orderly process. When disulfide bridges A and B are reduced, the proteins can be reoxidized to catalytically active proteins. However, reduction of the third disulfide bridge C requires more vigorous conditions, and enzyme activity cannot be recovered upon reoxidation.

3.3 Pepsin as an Esterase.

Until 1960 pepsin was thought to have a very narrow specificity. "Pepsin hydrolyzes only peptide linkages; it is not an esterase and will not attack amide linkages. Amino acid residues must be of the L-configuration on both sides of the peptide bond." [1]. However this narrow specificity of pepsin had to be broadened with the report in 1964 that pepsin hydrolyzes the ester acetyl-L-phenylalanyl $\nsim$ L-β-phenyllactic acid [131]. Inouye and Fruton [132] subsequently observed that the depsipeptide benzyloxyl-carbonyl-L-histidyl-p-nitro-L-phenylalanyl-L-β-phenyllactic acid methyl ester (Z-His-Phe(NO$_2$) $\nsim$ Pla-OMe) is rapidly cleaved by pepsin at the Phe(NO$_2$) $\nsim$ Pla bond. At pH 4, this ester exhibits a Michaelis constant very similar to that of the corresponding substrate (Z-His-Phe(NO$_2$) $\nsim$ Phe-OMe), but a three times larger rate constant. Compounds that are resistant to pepsin cleavage (cf. Table 5) act as competitive inhibitors of the enzymatic hydrolysis of both the ester and aforementioned peptide. The observation of identical inhibition constants provides support for the theory that both the ester and peptide are cleaved at the same active site. An advantage of working with this ester is that the kinetics of hydrolysis can be easily followed spectrophotometrically at 310 nm. Surprisingly, little additional research has been reported on the pepsin catalyzed hydrolysis of this ester and none on other similar esters. Further studies in this area of pepsin specificity should contribute important information toward the mechanism of pepsin action.

Shortly after it was established that pepsin hydrolyzes L-β-phenyllactic acid esters, Reid and Fahrney [133] discovered that organic sulfite esters (phenyl methyl sulfite and diphenyl sulfite) react very rapidly with pepsin. Rates were conveniently measured on the pH stat or spectrophotometrically by following the release of phenol. Z-Phe-Tyr acts as a competitive inhibitor for this reaction with a K_I equal to K_m, and N-diazoacetyl-norleucine methyl ester inhibits pepsin's reaction with these sulfite esters by better than 99%. Therefore, sulfite esters presumably are hydrolyzed by pepsin at the same active site as used for peptides.

Stein and Fahrney [135] using pepsin enriched with ^{18}O, presumably in an aspartic acid carboxyl group on the enzyme, showed that the bisulfite liberated in this reaction contained an excess of ^{18}O. The following mechanism was suggested to account for these results (eq. 9).

$$\text{Pepsin—C}^{18}\text{O}_2^{\ominus} + \text{ArO—}\overset{\overset{\displaystyle O}{\|}}{\text{S}}\text{—OR} \underset{K_m}{\rightleftharpoons} \text{pepsin—C}^{18}\text{O}_2^{\ominus}\cdot\text{ArO}\overset{\overset{\displaystyle O}{\|}}{\text{S}}\text{—OR}$$

$$\text{Pepsin—C}^{18}\text{O}_2^{\ominus}\cdot\text{Ar—O—}\overset{\overset{\displaystyle O}{\|}}{\text{S}}\text{—OR} \longrightarrow \underset{\text{I}}{\text{pepsin—}\overset{\overset{\displaystyle O}{\|}}{\text{C}}\text{—}^{18}\text{O—}\overset{\overset{\displaystyle O}{\|}}{\text{S}}\text{—OR}} + \text{ArOH} \quad (9)$$

$$\underset{\text{I}}{\text{Pepsin—}\overset{\overset{\displaystyle O}{\|}}{\text{C}}\text{—}^{18}\text{O—}\overset{\overset{\displaystyle O}{\|}}{\text{S}}\text{—OR}} \overset{\text{H}_2\text{O}}{\longrightarrow} \text{pepsin—CO}_2^{\ominus} + \text{HS}^{18}\text{O}_3^- + \text{ROH}$$

This reaction is identical to that in equation 1, with the covalently bound intermediate (E-Y) being a sulfite-pepsin anhydride. Further evidence that the reaction of pepsin with sulfite esters involves more than one step has recently been described by Hubbard and Stein [137], using bis-p-nitrophenyl sulfite. The kinetics of this reaction (measured in a stopped-flow instrument under conditions of (E) > (S) appears to be biphasic with an initial rapid release of p-nitrophenol followed by a slower release of a smaller amount. However, quantitative apportionment of p-nitrophenol to each step was not possible, and only an approximate rate constant could be calculated for the second step. Therefore, the use of these kinetic data to further support a pepsin-sulfite anhydride intermediate must await additional studies. In contrast, May and Kaiser using the same sulfite under similar experimental conditions observed only monophasic kinetics [136, 138].

May and Kaiser [136, 138] were the first to use bis-p-nitrophenyl sulfite (BNPS) for kinetic studies on pepsin. This is the most active substrate for pepsin yet found. For comparison at pH 1.91, the k_{cat} and K_m are 56.9 sec^{-1} and 0.25 mM for BNPS and 0.16 sec^{-1} and 30 mM, respectively, for methyl phenyl sulfite at pH 2.0 and 25°. A good peptide substrate (Table 3) for pepsin would have a comparable K_m but a much smaller rate constant. There are a number of unresolved questions involving sulfite esters, and this general area is certainly deserving of additional work. As one example, the pepsin-sulfite anhydride intermediate proposed in the foregoing equation should be capable of being trapped as has been done in the transpeptidation reactions (cf. section on trapping experiments).

As a practical application of this reaction, pepsin has been used to effect the first resolution of a sulfite ester [134]. Because of the asymmetry about the pyramidal sulfur atom, phenyl tetrahydrofurfuryl sulfite is capable of existing in two enantiomeric forms. Pepsin hydrolyzes only 50% of the racemic ester. The unreacted ester can then be extracted and is optically active. The enzymatic hydrolysis of methyl phenylsulfite proceeds to 100% completion, presumably because the small methyl group cannot promote stereochemical discrimination.

3.4 pH Dependence.

The pH dependency of enzyme catalyzed reactions is indicative of the number and type of ionizable catalytic groups at the active site. Pepsin exhibits bell-shaped pH-activity profiles [1, 67], but due to the experimental conditions under which these studies were carried out, it was difficult to analyze quantitatively these profiles. It was not until 1966 that the first pH-rate profile appeared [86]. This was for the pepsin catalyzed hydrolysis of Ac-Phe $\rightarrow$ Tyr-OMe, a neutral substrate in the pH range of pepsin activity (number 4, Table 10). Since then a number of pH-dependency studies have appeared, and the results are listed in Table X.

The simplest analysis of the pK_a's shown in Table 10 gives rise to the following pH dependency scheme for pepsin (equation 10):

$$
\begin{array}{ccc}
\mathrm{EH_2} & & \mathrm{EH_2 \cdot S} \\
\Big\updownarrow {\scriptstyle K_{E1}} & & \Big\updownarrow {\scriptstyle K_{ES1}} \\
\mathrm{EH + S} & \rightleftharpoons \mathrm{EH \cdot S} \xrightarrow{k_{cat}} & \text{products} \\
\Big\updownarrow {\scriptstyle K_{E2}} & & \Big\updownarrow {\scriptstyle K_{ES2}} \\
\mathrm{E} & & \mathrm{E \cdot S}
\end{array}
\tag{10}
$$

TABLE 10 pK_a's ASSOCIATED WITH THE CATALYTIC GROUPS CONTROLLING PEPSIN-CATALYZED HYDROLYSES

Substrate	pK_{E_1}	pK_{E_2}	pK_{ES_1}	pK_{ES_2}	pH of Maximum Activity for		Reference
					k_{cat}/K_m	k_{cat}	
1. Ac-Phe—Tyr-NH$_2$[a]	1.17	4.35	1.35	4.15	2.9	2.8	75
2. AcPhe—Tyr[a]	1.17	—	1.12	3.70	2.3	2.3	75
3. Ac-Phe—Trp[a]	1.40	—	1.05	3.70	2.3	2.4	75
4. Ac-Phe—Tyr-OMe[b]	—	—	1.6 (1.6)	3.5 (4.1)	—	2.7 (3)	23, 86
5. Ac-Phe—Tyr-OMe[c]	—	—	2.1	4.6	2.7	3.3	27
6. Ac-Phe—Tyr-OMe[d]	<1	≈4.5	—	—	2.7	—	94
7. Ac-Phe—Phe-OMe[b]	—	—	(1.4)	(4.6)	—	(3.1)	23
8. Ac-Tyr—Phe-OMe[b]	—	—	(1.8)	(4.6)	—	(3.0)	23
9. Ac-Phe—I$_2$Tyr	<1	3.2	—	—	2.0	—	58
10. AcPhe—Br$_2$Tyr	0.75	2.67	0.89	3.44	1.9	2.2	74
11. Ac-Phe—Phe[e]	1.1	3.5	—	—	2.3	—	139
12. Ac-Phe—Phe-Gly[e]	1.1	3.5	—	—	2.2	—	139
13. Ac-Phe—Phe-NH$_2$[f]	1.05	4.75	—	—	2.9	—	139
14. Z-His—Phe-OEt	≈3.7	>5	≈3.5	≈5.2	3.9	4.5	60
15. Methylphenyl sulfite[g]	—	—	2.6	—	—	—	133
16. bis p-NO$_2$ phenyl sulfite[h]	0.82	5.17	0.57 (1.35)	4.73	3	2.8	138

[a] Solvent 3% methanol; 35°.
[b] Solvent 3% dioxane; 25°. The pK's in parantheses were measured using pepsin obtained from the activation of pepsinogen.
[c] Solvent 3% dioxane; 25°. The pK's were measured using pepsin obtained from the activation of dephosphorylated pepsinogen. The phosphate group in pepsinogen was removed enzymatically using potato monophosphotase.
[d] Solvent water; 37°.
[e] Solvent water; 37°.
[f] Solvent 1.2% dimethylformamide; 37°.
[g] Solvent water; 25°. Kinetic constants were not determined above pH 4.0.
[h] Solvent 0.4% acetonitrile; 25°.

This scheme implies that:

1. There are two catalytically important ionizing groups in the free enzyme.

2. There are two catalytically important groups on pepsin participating in the catalytic step.

3. Only EH binds to the substrate.

4. Only the EH·S (pepsin-substrate complex) reacts to give product.

5. All ionizations occur faster than either pepsin-substrate binding or catalysis.

6. The state of ionization of the substrate is the same throughout this pH range. This is not true for all substrates listed in Table 10.

The pK_{E_1} and pK_{E_2} are calculated from equation 11, which describes a bell-shaped pH dependency of the catalytic constant (k_{cat}/K_m). The pH dependency for this catalytic constant describes the catalytically important ionizations of the free pepsin [140].

$$\frac{k_{cat}}{K_m} = \frac{(k_{cat}/K_m)\text{limit}}{1 + \dfrac{H}{K_{E_1}} + \dfrac{K_{E_2}}{H}} \tag{11}$$

This bell-shaped pH dependency of k_{cat}/K_m is exhibited by the neutral compounds 1, 6, 13, 14, and 16 in Table 10. Table 10 shows a large number of pK_{E_1} values that appear to be similar and therefore thought to be ascribable to the same group. For pK_{E_2} there are considerably less available data because the ionization of the free carboxyl group in compounds 2, 3, 9, 10, 11, and 12 in Table 10 has to be included in the calculation of pK_{E_2}. The pK_a of the free carboxyl group in a substrate such as Ac-Phe-Tyr is about 3.5 [74, 75]. In fact, for substrates with a free carboxyl group capable of ionizing in this pH range, the anionic form of the substrate is thought to be unreactive toward pepsin [75]. Undoubtedly, for compounds 2, 3, 9, 10, 11, and 12 in Table 10 the pK_{E_2} listed does not reflect the ionization of a group on the free enzyme alone but also the ionization of the carboxyl group in the substrate.

The kinetic measurement of k_{cat}/K_m offers two additional advantages. First, it is easily determined for substrates that are sparingly soluble in aqueous buffer solutions [139]. According to the Michaelis-Menten equation, the kinetic determination of a separate k_{cat} and K_m requires a substrate that exhibits a solubility at least equal to K_m. Second, it can be shown for reactions involving more than one catalytic step that k_{cat}/K_m is equivalent to k_2/K_s. The k_2 is the rate constant for the first catalytic step and K_s is the true enzyme substrate binding constant free of any influence by rate constants.

The pK_{ES_1} and pK_{ES_2} are obtained from equation 12, which has been applied to substrates exhibiting a bell-shaped dependency for hydrolysis by pepsin.

$$k_{cat} = \frac{(k_{cat})\text{limit}}{1 + \dfrac{H}{K_{ES_1}} + \dfrac{K_{ES_2}}{H}} \tag{12}$$

Again the neutral substrates, compounds 1, 4, 7, 8, 14, and 16 in Table 10, which do not ionize in this pH region, yield the most unambiguous pK_a's. It should be noted that k_{cat} could be a combination of rate constants. This would yield a composite pH-k_{cat} profile reflecting the proportionate pH dependency of each individual rate constant contributing to k_{cat}. However, evidence has been presented in the sections on specificity and inhibition of pepsin catalyzed reactions that *all* synthetic substrates used for pepsin are hydrolyzed with a k_2 slow step. That is, the slow step for hydrolysis immediately follows the formation of the Michaelis complex. This general assumption does bear a further reevaluation.

The pH-dependency studies to date suggest that there are two catalytically important groups on the free enzyme and that the same two groups control the reaction of the enzyme-substrate complex. The pH for maximum activity of pepsin with the substrates listed in Table 10 is approximately 3. The magnitude of the pK_a's of the catalytically important groups suggests that one is a protonated carboxyl group and the other a carboxylate, acting in a concerted manner. The similarity of pK_{E_1} with pK_{ES_1} suggests that these are dissociation constants of the same carboxyl group, with small shifts being caused by substrate binding. The same conclusion can be made for pK_{E_2} and pK_{ES_2}.

Two possible alternatives for the identity of these catalytically important groups are: that the pK_a's around 1 involve a protonated amide or peptide bond [23] or that the bell-shaped dependencies are due to the participation by a tetrahedral intermediate as observed for the hydrolysis of O-carboxyphthalimide [141]. Neither alternative can be excluded for the time being, but it is quite unlikely that they would provide the key to the understanding of the mechanism of pepsin action.

That the phosphate group in pepsin *does not participate* in the mechanism of action of pepsin was established by measuring the kinetics and pH dependency of the dephosphorylated pepsin (pepsin D) catalyzed hydrolysis of Ac-Phe-Tyr-OMe [27]. The pH dependency of the pH-k_{cat} curve is bell-shaped and very similar to that obtained with pepsin A (compare compounds 4 and 5, Table 10). A comparison of the k_{cat} (limit) for the two hydrolyses shows that the dephosphorylated pepsin is 89% as active as pepsin A. It is now accepted that this phosphate group is in the monoester form in either pepsin A or pepsinogen A [27].

The above discussion on the pH dependency of pepsin catalyzed hydrolyses forms the basis of a convenient interpretation that can be readily incorporated into a proposal on the mechanism of pepsin action. However, this review would be incomplete if we did not point out some of the inconsistencies obtained in some of the pH dependency studies:

1. An examination of the pK_{E_1} from Table 10 shows quite a variation in magnitude with substrate. If this pK_{E_1} is indeed a measure of the pK_a of the same carboxyl group on the free pepsin, there is no readily available explanation to account for the observed variation with substrate.

2. Neutral dipeptide amides (e.g., Ac-Phe $\rightarrow$ Tyr-NH$_2$, number 1, Table 10) yield symmetrical bell-shaped pH-k_{cat} and pH-k_{cat}/K_m profiles [75]. However, the corresponding neutral methyl ester (number 4, Table 10, Ac-Phe $\rightarrow$ Tyr-OMe) exhibits a bell-shaped pH-k_{cat} dependency but a very poor bell-shaped pH-k_{cat}/K_m profile. It is indeed surprising that this simple change from amide to methyl ester at a position quite removed from the susceptible bond could cause such a dramatic change in the pH-k_{cat}/K_m profile. The change from commercial pepsin to pepsin obtained by activating pepsinogen produces flat, pH independent k_{cat}/K_m curves for the neutral methyl esters, numbers 4, 7, and 8 in Table 10.

The measured K_m's are larger for hydrolyses using pepsin from pepsinogen rather than commercial pepsin [23]. Unfortunately, the three methyl esters (numbers 4, 7, and 8, Table 10) are the only pepsin substrates for which the pH dependency has been determined with pure pepsin. The kinetic behavior of the corresponding dipeptide amide should also be studied with pure enzyme. No ready explanation for this change in kinetic behavior with pepsin preparation is currently available.

3. May and Kaiser [138] observed bell-shaped pH-k_{cat} and k_{cat}/K_m profiles for the pepsin catalyzed hydrolysis of bis-p-nitrophenylsulfite (number 16, Table 10). Except for pK_{ES_1}, the other pK_a's are in good agreement with those found with the more specific peptide substrates (cf. Table 10). This result is not surprising because evidence was presented in the section on esterase activity that sulfite esters and peptides react at the same active site on pepsin. However, the two pK_{ES_1} values indicate that the acidic side of the pH-k_{cat} profile can not be analyzed on the basis of a simple bell-shaped profile. The extremely low pK_{ES_1} value of 0.57 was calculated assuming a symmetrical curve. A more comparable value of 1.35 was calculated, assuming a leveling in the lefthand side of the pH-k_{cat} profile. The simplest explanation for this behavior is that not only does EHS react kinetically to give products but so does EH$_2$·S. The possibility that more

than one ionized form of the enzyme-substrate complex can react kinetically could be considered for future pH-dependency studies on pepsin.

4. The pH-dependency curves of the cationic substrates of the type Z-His-Phe $\leftthreetimes$ Phe-OEt (number 14, Table 10) present still further problems of interpretation. The solubility of these compounds above pH 4 is so poor as to prevent kinetic measurements of k_{cat} and K_m over the entire pH range, 1–6, of pepsin activity. Often, only the pH-k_{cat}/K_m (pH activity) profiles have been reported for these substrates. The one cationic substrate studied in some detail, Z-His-Phe $\leftthreetimes$ Phe-OEt (number 14, Table 10), exhibits unsymmetrical bell-shaped pH-k_{cat} and k_{cat}/K_m curves with pK_{E_1} and pK_{ES_1} of 3.5 and 3.7. These are considerably larger than the same pK_a's reported for other peptide substrates (Table 10). The possibility of a change in pK_a of the imidazolium group in the substrate on binding to pepsin must also be considered. If this group does ionize in the pH range of interest, the pH-dependency profiles would surely reflect this ionization. The pH-k_{cat}/K_m profiles for Gly-Gly-Phe $\leftthreetimes$ Phe-OEt and Z-His-Tyr $\leftthreetimes$ Tyr-OEt are also unsymmetrical bell-shaped curves with pK_{E_1} of about 3.5. Furthermore for Bz-Lys-Phe $\leftthreetimes$ Phe-OEt, Gly-Gly-Phe $\leftthreetimes$ Phe-OEt and isonicotinoyl-Phe $\leftthreetimes$ Phe-OP4P the k_{cat} for hydrolysis of the Phe $\leftthreetimes$ Phe bond is relatively constant in the pH 2–4.5 range, with the increase in k_{cat}/K_m reflecting a decrease in K_m (79, 83).

These discrepancies indicate that the details of the pH dependency of pepsin action are far from being firmly established. Moreover, the diazonium compounds inactivated a carboxyl group of an aspartic acid with a pK_a of about 5–5.5. This pK_a is a full one unit higher than the pK_{ES_2}'s implicated in the hydrolysis reactions. This has led some workers in the field [60, 108] to suggest that the active site of pepsin contains a series of carboxyl groups with overlapping pK_a's whose participation in binding, catalysis, and irreversible inhibition depends on the nature of the substrate. Although this type of pH behavior has not been observed with other proteolytic enzymes, it should not be ignored or dismissed for pepsin.

3.5 Deuterium Oxide Solvent Isotope Effects

The effect of D_2O on the catalytic rate of enzyme-catalyzed reactions has been used to distinguish between general base and nucleophilic mechanisms. A kinetic solvent isotope of 2–3 observed for α-chymotrypsin is consistent with rate-limiting proton transfer. A D_2O kinetic isotope effect of this magnitude is often used as evidence that a proton transfer is part of the rate-limiting step. There have been three reported measurements of the D_2O kinetic solvent isotope for pepsin catalyzed hydrolyses.

First, Clement and Snyder [86] compared the bell-shaped pH-k_{cat} profile for the pepsin catalyzed hydrolysis of Ac-Phe $\rightarrow$ Tyr-OMe carried out in H_2O with that done in D_2O. There was no observed D_2O effect, $k_{cat}(H_2O)/k_{cat}(D_2O) = 1.05 \pm 0.30$. The expected shift in pK_{ES_1} $1.6 \rightarrow 1.9$ and pK_{ES_2} $3.5 \rightarrow 4.0$ was observed with the solvent change from H_2O to D_2O. Second, Reid and Fahrney [133] reported similar results, $k_{cat}(H_2O)/k_{cat}(D_2O) = 1.1 \pm 0.1$ for the pepsin catalyzed hydrolysis of methyl phenyl sulfite. The pH-k_{cat} profile is a sigmoid at pH values lower than 4.0, with a pK_{ES_1} of 2.6 in H_2O and 3.1 in D_2O. The positive shift of 0.3–0.50 pK_a units in going from water to deuterium oxide is a generally observed property of all acids. Third, Hollands and Fruton [143] compared the kinetics of the hydrolysis of Gly-Gly-Gly-Phe(NO_2) $\rightarrow$ Phe-OMe in H_2O at pH 4 with that found in D_2O at pD 4.4. For this *one* measurement there was a meaningful D_2O effect $(k_{cat})H_2O/(k_{cat})D_2O = 2$; $(k_{cat}/K_m)H_2O/(k_{cat}/K_m)D_2O = 1.7$.

The first two results suggest an absence of general acid-base catalysis in the rate-limiting step of pepsin hydrolysis of those two substrates. The magnitude of the catalytically important pK_a's would suggest a nucleophilic attack by a carboxylate anion as participating in the rate-limiting step. The observation of a D_2O effect in the third study on a cationic substrate suggests just the opposite interpretation, namely an enzymatic carboxyl group acting as a proton donor in the rate-limiting step. Clearly, additional D_2O solvent isotope studies are required before they can be applied as an unambiguous criteria for the mechanism of pepsin action. Because water is one of the substrates in all hydrolytic enzymatic reactions, the absence of other secondary kinetic solvent isotope effects [144] should be established in each case.

The mechanistic implications of the presence or absence of a kinetic deuterium oxide solvent isotope must be carefully interpreted, but they can be used successfully to produce one part of the total package of information that forms the basis of an enzyme mechanism. We believe that the absence of a D_2O effect for the pepsin catalyzed hydrolysis of two substrates is a mechanistically meaningful observation. The presence or absence of a D_2O effect that depends on the substrate is usually suggestive of different mechanisms. Therefore, Ac-Phe-Tyr-OMe and methyl phenyl sulfite are hydrolyzed by pepsin with one rate-limiting step, whereas for Gly-Gly-Gly-Phe(NO_2)-Phe-OMe another step must be rate limiting. This explanation is entirely consistent with the mechanistic picture of pepsin action presented so far involving more than one catalytic step. This suggestion does mandate a reevaluation of some of the binding data (section on inhibition), which suggests that the slow step for all pepsin catalyzed hydrolyses immediately follows the formation of the Michaelis complex.

For the carboxypeptidase catalyzed hydrolysis of N-(N-benzoylglycyl)-L-phenylalaninate, there is no D_2O effect on k_{cat}/K_m, and that for k_{cat} is small

$((k_{\rm cat})\,{\rm H_2O}/(k_{\rm cat})\,{\rm D_2O}) = 1.33)$. However, the $(k_{\rm cat})\,{\rm H_2O}/(k_{\rm cat})\,{\rm D_2O}$ ratio for the carboxypeptidase catalyzed hydrolysis of O-($trans$-cinnamoyl)-L-β-phenyl-lactate is 2 [145, 146]. Kaiser and Kaiser [145, 146] used these data to suggest that carboxypeptidase hydrolyzes esters and peptides with different rate-limiting steps. For peptides the formation of an anhydride is rate determining (no proton transfer required), but for specific esters the breakdown of this anhydride involving a proton transfer is rate determining. This same type of approach could also apply to pepsin. Additional D_2O kinetic solvent isotope effect studies on pepsin catalyzed hydrolyses of Ac-Phe $\prec$ Tyr-NH$_2$ and bis-p-nitrophenyl sulfite should produce some mechanistically important data. These are the two substrates that exhibit both bell-shaped pH-$k_{\rm cat}$ and $k_{\rm cat}/K_m$ dependencies in H_2O.

4 EVIDENCE FOR PEPSIN-SUBSTRATE INTERMEDIATES

4.1 Introduction

Proteolytic enzymes are known to be efficient catalysts for both hydrolyses and transfer reactions [3]. In hydrolysis reactions, water acts as the acceptor molecule, whereas in transfer reactions other organic molecules fulfill this role according to equation 13. The E-P$_2$ is an enzyme substrate covalently

$$ {\rm E} + {\rm P_1\text{-}P_2} \rightleftharpoons {\rm E\cdot P_1\text{-}P_2} \xrightarrow[-P_1]{} {\rm E\text{-}P_2} \xrightarrow[\text{hydrolysis}]{{\rm H_2O}} {\rm E} + {\rm P_2} $$
$$ \Big\downarrow{\scriptstyle N} \atop \xrightarrow[\text{transfer}]{} {\rm N\text{-}P_2} \tag{13} $$

bound intermediate that partitions itself between water and an acceptor, N. If N-P$_2$ is stable under the reaction conditions, the enzyme thus catalyzes the synthesis of another material. In addition the stereochemical constraints imposed by the enzyme should lead to stereoselective syntheses. The most familiar transfer enzymes are the serine and cysteine proteases, which can act as general acyl transferases,

$$ \underset{\text{donor}}{\text{Ac-Tyr} \prec \text{Tyr}} + \underset{\text{acceptor}}{\text{E-COOH}} \longrightarrow \text{Ac-Tyr} + \overset{\displaystyle O}{\overset{\|}{\text{E-C}}}\text{-Tyr} $$

$$ \underset{\text{donor}}{\overset{\displaystyle O}{\overset{\|}{\text{E-C}}}\text{-Tyr}} + \underset{\text{acceptor}}{\text{Ac-Tyr-Tyr}} \longrightarrow \text{E-COOH} + \text{Ac-Tyr-Tyr-Tyr} \tag{14} $$

$$ \text{Ac-Tyr} \prec \text{Tyr-Tyr} + \text{E-COOH} \longrightarrow \text{AcTyr} + \text{Tyr-Tyr} + \text{E-COOH} $$

Pepsin can also catalyze transfer reactions. Neumann et al. [147] observed that the pepsin catalyzed hydrolysis of Ac-Tyr-Tyr, N-p-aminobenzoyl-Tyr-Tyr, Z-Glu-Tyr and Z-Phe-Tyr all produced detectable quantities of Tyr-Tyr along with the expected hydrolysis products Ac-Tyr and Tyr. Considering the known specificity of pepsin, a transpeptidation (amine transfer) reaction was proposed involving an amino-pepsin covalently bound intermediate. The reaction is best described by equation 14.

$$\overset{O}{\overset{\|}{}}$$

The second reaction is transpeptidation and E-C-Tyr is the amino-pepsin covalently bound intermediate:

$$E-\overset{O}{\overset{\|}{C}}-NH-\overset{H}{\underset{\underset{\underset{OH}{\bigcirc}}{CH_2}}{C}}-COOH$$

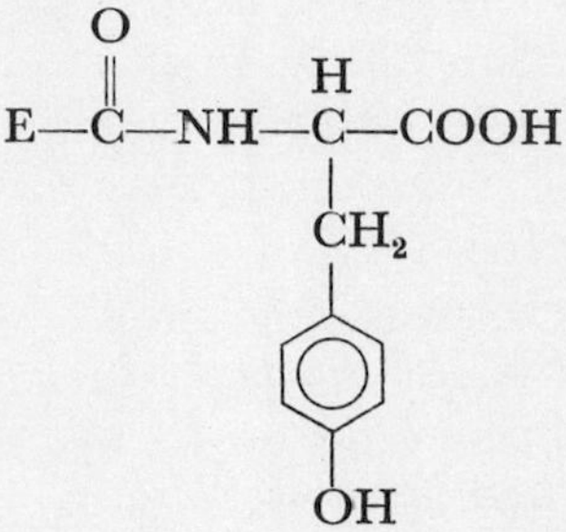

On the basis of the large amount of data implicating a carboxyl in the mechanism of pepsin action, this group has been proposed as the amino transfer group in the active site of pepsin.

Sharon et al. [148] reported that pepsin catalyzes the exchange of ^{18}O from water into the virtual substrate Z-Phe. This reaction was shown to be specific since ^{18}O was not incorporated into the corresponding D-isomer. The similar isotope exchange reactions catalyzed by the proteolytic enzymes chymotrypsin [149] and papain [150] suggest that this reaction proceeds through an intermediate acylenzyme. For pepsin, a reasonable acyl-transfer group in the active site would be a carboxyl. The ^{18}O and transpeptidation reactions support the postulate [23, 148] that both the acyl and imino groups of the susceptible peptide bond become covalently linked to pepsin (eq. 15).

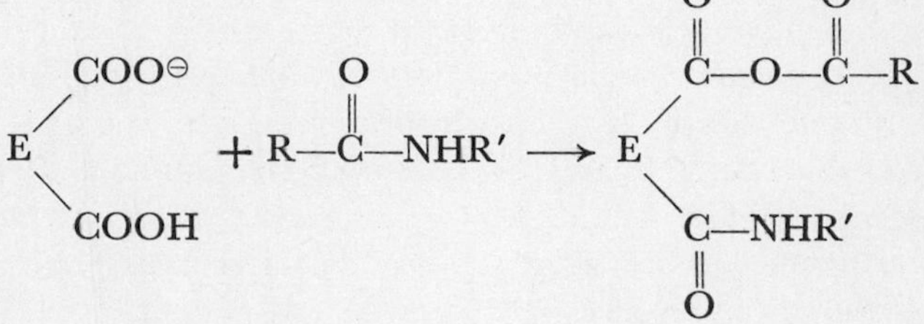

$$E\overset{\displaystyle COO^{\ominus}}{\underset{\displaystyle COOH}{\big<}} + R-\overset{O}{\overset{\|}{C}}-NHR' \longrightarrow E\overset{\displaystyle \overset{O}{\overset{\|}{C}}-O-\overset{O}{\overset{\|}{C}}-R}{\underset{\displaystyle \underset{O}{\underset{\|}{C}}-NHR'}{\big<}} \qquad (15)$$

As the following discussion shows, the amino-pepsin intermediate is reasonably well established, but the presence of the anhydride linkage involving the acyl group is still questionable. Two papers [147, 148] have formed the foundation for considerable further research on pepsin-substrate intermediates. The following discussion is divided into three sections: (1) kinetic and trapping reactions, (2) additional transpeptidation, and (3) further ^{18}O exchange studies.

4.2 Kinetic and Trapping Experiments

If pepsin-substrate intermediates form during pepsin catalyzed hydrolyses, they should be kinetically detectable if their rate of breakdown is slower than their rate of formation. The detection of presteady state or "burst" kinetics requires the following experimental conditions: $(S_0) > K_m$ and $k_2/k_3 \gg 1$ (where k_2 is the rate constant for formation of the intermediate and k_3 that for the breakdown); (E_0) must be high enough to detect a significant amount of intermediate; whereas the condition $(S_0) > (E_0)$ must be met in order to observe a subsequent steady state reaction [50].

For pepsin these experimental conditions are difficult to meet because of low substrate solubilities and generally large K_m's. Cornish-Bowden et al. [90] did not observe a burst of phenylalanine by following the production of ninhydrin positive material [59, 61] during the pepsin hydrolysis of Ac-3,5-dinitro-Tyr —⟍ Phe. The detection of a burst of the acyl portion acetyl dinitrotyrosine of the substrate was thwarted by the lack of sensitivity of the spectrophotometric method (differential molar absorbtivity of 300). Inouye and Fruton [64] were unable to detect a burst of the acyl moiety for the pepsin hydrolysis of Ac-His-Phe(NO$_2$) —⟍ Phe-OMe. The progress of this reaction was followed at 310 nm with a higher differential molar absorbtivity of 800. These two results would suggest that the formation of pepsin-substrate intermediates is the slow step in these hydrolyses. However, for presteady state kinetics to be observed, much more sensitive pepsin substrates must be developed. To date, the restricted specificity requirements of pepsin have thwarted all attempts to design such substrates.

May and Kaiser [151] have extended their studies on bis-p-nitrophenyl sulfite. They find that for the pepsin catalyzed hydrolysis of this sulfite ester at pH 1.5 both p-nitrophenol and inorganic sulfite are liberated concurrently. Therefore, if pepsin intermediates are formed during the hydrolysis of this sulfite ester, the slow step must be their formation. This report is in conflict with that of Hubbard and Stein [137] who observed biphasic kinetics for this same reaction at pH 2.2. The evidence that pepsin-substrate intermediates do occur during the hydrolysis of some sulfite esters was presented in the section on pepsin as an esterase.

Evidence *against* an acyl-enzyme anhydride intermediate has been obtained from trapping experiments. Cornish-Bowden et al. [90] were unable to detect any Ac-Phe-OMe produced in the hydrolytic reaction of Ac-Phe $\rightthreetimes$ Phe-Gly or the virtual reaction of Ac-Phe in the presence of (^{14}C) methanol. Similar trapping experiments with hydroxylamine were also negative [152]. Of course, negative trapping results are never conclusive. It is possible that an acyl-pepsin anhydride intermediate is present but that the anhydride link is inaccessible to methanol or hydroxylamine. However methanol is more than able to compete with water in other hydrolytic enzymes such as α-chymotrypsin [153] and papain [154]. For a further discussion on this point refer to the section on oxygen exchange.

The report that when pepsin is incubated with Z-Tyr or Z-Tyr-Tyr in the presence of C^3H_3OH, radioactivity is incorporated into the enzyme has been withdrawn [155, 156]. Kitson and Knowles [157] were unable to substantiate this result. This means that if an acyl-anhydride intermediate (ECOOOCR) is part of the pepsin mechanism, neither carbonyl carbon is susceptible to nucleophilic attack by methanol. The complexity of the experimental part of many recent papers on the mechanism of pepsin has significantly increased. Considering the erroneous conclusions that could have been drawn from the above report, the mechanistic conclusions reached in these papers should be treated cautiously.

There are two reports of trapping amino pepsin intermediates. The incubation of pepsin with Z-Glu $\rightthreetimes$ (^{14}C)Tyr yields a radioactive pepsin upon precipitation from solution. When Z-(^{14}C)Glu-Tyr is used as the substrate no radioactivity remains associated with the enzyme [158]. Similar results were obtained when Z-Tyr(^{14}C)Tyr and Z-(^{14}C)Tyr-Tyr were the substrates [159]. Because of the unreproducibility of the C^3H_3OH labeling experiments, one author [156] suggests that his amino-trapping results be treated cautiously until the experiments have been reevaluated.

4.3 Transpeptidation

Considering the possible uncertainties in the trapping experiments, the transpeptidation reaction [147] represents the primary evidence for an amino-pepsin intermediate. This reaction was further verified by Fruton et al. [160] who incubated pepsin at pH 4 with two solutions of:

(a) Z-Tyr-Tyr; Z-(^{14}C)Tyr; and Tyr

(b) Z-Tyr-Tyr; Z-Tyr; and (^{14}C)Tyr

After termination of the reaction, product separation and removal of the benzyloxycarbonyl group, solution (a) showed incorporation of radioactive Tyr into the Tyr-Tyr; whereas the Tyr-Tyr from (b) was nonradioactive

The thermodynamics of this transpeptidation (condensation) reaction have been measured for the following pepsin catalyzed reaction at pH 4.0 [161].

$$\text{Ac-Phe} \rightleftharpoons \text{Tyr-OEt } H_2O \xrightarrow{\text{pepsin}} \text{Ac-Phe} + \text{Tyr-OEt} \qquad (16)$$

The equilibrium constant for hydrolysis is described by:

$$K_H = \frac{(\text{Ac-Phe})(\text{Tyr-OEt})}{(\text{Ac-Phe-Tyr-OEt})} = 1 \qquad (17)$$

This means that these condensation reactions are thermodynamically feasible. The free energy of hydrolysis of the peptide bond in Ac-Phe + Tyr-OEt is $\Delta G = O$ at 25° and a pH of 4. Interestingly, a synthesis of Ac-Phe-Tyr-OEt (9% yield), Ac-Tyr-Tyr-OEt (12% yield), and Ac-Phe-Phe-OEt (11% yield) was accomplished by incubating pepsin with the corresponding amino acids (i.e., AcPhe plus Tyr-OEt at pH 4.0 and 37°). A consideration of the known specificity of pepsin (it reacts only with L-amino acids) leads one to suggest that pepsin could be utilized for peptide synthesis. The synthesis of acyl dipeptide esters is even more surprising in light of recent reports [75, 162] that Ac-Phe-Trp-NH₂, Ac-Phe-Tyr-NH₂, and Ac-Phe-Phe-OEt do not participate in a transpeptidation reaction. Neither Ac-Phe-Tyr-NH₂ nor Ac-Phe-Phe-OEt yields an intermediate that could be trapped by Ac-(^{14}C)Phe. Also, Clement et al. [23, 27] observed no transpeptidation reaction during the pepsin catalyzed hydrolysis of Ac-Phe-Tyr-OMe, Ac-Phe-Phe-OMe, and Ac-Tyr-Phe-OMe as evidenced by quantitative hydrolysis obeying simple kinetics. For these *neutral dipeptides* transpeptidation seems to be absent over the entire pH range of pepsin activity.

For the acyl peptide substrates containing a *free C*-terminal carboxyl group, the transpeptidation reaction is appreciable at pH values > 4 but nonexistent or only slightly detectable at pH's < 2. Kitson and Knowles [92a] found the following transpeptidation reaction to be significant at

$$\text{Ac-(^{3}H)Phe} + \text{Ac-Phe-Phe-Gly} \xrightarrow{\text{pepsin}} \text{Ac-Phe} + \text{Ac-(^{3}H)Phe-Phe-Gly}$$
$$(8.5\% \text{ at } 85\% \text{ hydrolysis}) \qquad (18)$$

pH 4.7 but undetectable at pH 1.3. However, they have not established the proportion of transpeptidation product which might have been formed by substrate-product equilibrium. Ac-Phe $\rightleftharpoons$ Tyr and Ac-Phe $\rightleftharpoons$ Trp yield transpeptidation products (Tyr-Tyr and Trp-Trp, respectively) at pH's > 4 but almost none at pH 2 [75, 94].

Mal'tsev et al. [163] investigated the specificity of pepsin in these transpeptidation reactions at pH 4.7 and 37° as exemplified by the following reaction:

$$\text{Ac-Phe-Tyr} + \text{Z-Phe} \xrightleftharpoons{\text{pepsin}} \text{Ac-Phe} + \text{Z-Phe-Tyr} \qquad (19)$$

donor   acceptor   transpeptidation product

The specificity of pepsin as measured by the rate of the transpeptidation reaction with respect to the acceptor molecule is Z-Phe > Z-Tyr > Z-Leu > Z-Glu. No transpeptidation product was found with the acceptors Z-Ala, Z-Ser, Z-Gly, and Z-Val. The specificity of the acceptor molecule is predictable and consistent with pepsins' known specificity requirements for hydrolysis. In addition the rate of formation of transpeptidation product was correlated with the rate of hydrolysis of the peptide donors: Ac-Phe-Tyr > Ac-Phe-Phe > Ac-Phe-Leu. Using this observation as supporting evidence, Silver and Stoddard [162] present data consistent with the hypothesis that hydrolysis and transpeptidation when it occurs share a common rate-determining step.

The report [162] that the pepsin catalyzed hydrolysis of Ac-Phe-Tyr and Ac-Phe-Phe at pH 4.5 in the presence of Ac(^{3}H)Phe produces Ac-(^{3}H)Phe-Tyr and Ac-(^{3}H)Phe-Phe, respectively, whereas neither Ac-Phe-Tyr-NH$_2$ nor Ac-Phe-Phe-OEt yields an amino enzyme that Ac(^{3}H)Phe can trap is rather disconcerting. The evidence to date suggests that the transpeptidation reaction can be detected only for substrates having a free carboxyl group and then only at pH's > 4. Silver [162] has suggested that "Mechanistic generalizations about pepsin founded on the transpeptidation reaction certainly require more experimental justification . . . . Dropping the amino-enzyme as an intermediate directly on the hydrolysis pathway has some appeal."

A detailed listing of their complex arguments is clearly beyond the scope of this chapter. However, they do believe that enzymatic discrimination between E-Tyr-COOH and E-Tyr-CONH$_2$ cannot account for this difference in reactivity, where E-Tyr-COOH transfers a tyrosine group to an acceptor molecule but E-Tyr-CONH$_2$ does not. Attempts to salvage the amino-enzyme intermediate by expanding previously proposed mechanisms were unsuccessful. There is a report that one pepsin preparation showed no transpeptidation activity [164]. However, this one observation does require further experimental justification before it can be meaningfully incorporated into a general mechanism for pepsin action.

We would like to suggest that specificity arguments can indeed account for the above difference in the two amino transfer reactions. It is obvious that the specificity of the enzyme can be different for the transpeptidation reaction and for the hydrolysis reaction. Two reaction pathways that are the microscopic reverse of one another need not show the same specificity. For α-chymotrypsin, the specificity of the enzyme is different for acylation and for deacylation even though they are the microscopic reverse of one another [165]. The rate-limiting steps of the forward and the backward reaction are not necessarily the same. The data of Silver and Stoddard [162] might suggest that the hydrolysis of the peptide bond is *not* a one-step reaction and that each step has its own specificity. Obviously, more research must be done before one can say definitely that the amino enzyme intermediate is or is not a mandatory species in the pathway of pepsin catalyzed hydrolysis.

The pH dependency of the transpeptidation reaction [92, 162] is consistent with the proposal that the acceptor molecule (i.e., Ac-Phe) must be in its ionized, carboxylate form. Interestingly, the inhibition constant (Table 6) for Ac-Phe-COO$^\ominus$ is significantly larger than for Ac-Phe-COOH. This means that the anionic acceptor molecule does not bind as strongly to the enzyme as does its unionized form. Considering that at pH 4 one half of pepsin's 40 carboxyl groups are ionized [166] it is surprising that the enzyme can bind a negatively charged molecule at all. In the pH range where transpeptidation is important, the pepsin molecule carries a large negative charge. In addition, the active site also includes an ionized carboxyl con-

tributed by the amino enzyme donor molecule (i.e., E-C-NH-Tyr-COO$^\ominus$ with O double bonded to C). If the amino enzyme is a true intermediate in all pepsin catalyzed reactions, it is not clear at this time how the presence of this anionic group promotes the transpeptidation reaction.

4.4 Further ^{18}O Studies

Since the original report by Sharon et al. [148], a number of studies have recently appeared on the pepsin catalyzed ^{18}O exchange into virtual substrates. Kozlov et al. [167] found a parallelism between the magnitude of ^{18}O exchange from the solution into the carboxyl group of an acyl-amino acid and capacity of the carboxyl group for participating as an acceptor in the transpeptidation reaction. This concept is supported by the observation that free amino acids neither participate in ^{18}O exchange nor act as acceptors for the transpeptidation reaction [160, 167].

Kozlov et al. [168] measured the quantity of exchange between H$_2^{18}$O and added Ac-**Phe**-COOH in the presence and absence of the substrate, Ac-Phe-Tyr. The isotope exchange was 27%, and the presence of Ac-Phe-Tyr had no effect at pH 2.4. However, at pH 4.7 the added substrate increased the extent of exchange by 7% from 34.0 ± 1.0 to 41.3 ± 1.1%. This increased exchange is most easily described by equation 20, in which

$$\text{Ac-Phe-Tyr-COO}^\ominus + \text{Ac-\textbf{Phe}-COO}^\ominus \xrightarrow[\text{pepsin}]{\text{H}_2^{18}\text{O}} \text{Ac-\textbf{Phe}-Tyr}$$
$$+ \text{Ac-\textbf{Phe}-C}^{18}\text{OO}^\ominus \quad (20)$$

transpeptidation is important. Silver and Stoddard [162] used their trans-peptidation data to calculate the increased exchange expected in the presence of Ac-Phe-Tyr. These calculations showed that the exchange of $H_2{}^{18}O$ into Ac-Phe-COO$^\ominus$ should have been increased by only *one-tenth*, namely 0.75 % instead of the measured 7.3 %. This discrepancy cannot be reconciled with equation 20 and the simple amino-enzyme intermediate

$$\overset{\text{O}}{\overset{\|}{}}$$

proposal. Apparently E-C-NH-Tyr-COO$^\ominus$, the enzyme-tyrosine inter-mediate, catalyzes many exchanges of $H_2{}^{18}O$ with Ac-Phe-COO$^\ominus$ before returning to Ac-Phe-Tyr or proceeding to Tyr, the final product. Con-sidering the importance this discrepancy has on the generality of the amino-enzyme intermediate, additional ^{18}O exchange studies are clearly indicated. It should be pointed out that two of the most current pepsin mechanisms [144, 152] cannot be reconciled with this increased ^{18}O exchange being promoted by transpeptidation.

Kozlov et al. [169] examined the specificity displayed by pepsin in isotope exchange reactions of the oxygens of acetyl amino acids. The following order was observed for rate of ^{18}O incorporation into the carboxylate group of the acetyl amino acids at pH 4.0 and 19.5°: Ac-Phe > Ac-Nle > Ac-Leu > Ac-Trp > Ac-Nva > Ac-Tyr > Ac-Ala > Ac-Val. This series for ^{18}O exchange should be compared with similar series for acceptor effectiveness in the transpeptidation reaction [163] (refer to previous section on trans-peptidation). It is evident that there is close agreement between the two series. This is not an especially surprising correlation because both series parallel the known specificity requirements with respect to the X residue of the substrates A-X $\rightleftharpoons$ Y-B and Z-His-X-Phe-OMe (cf. Tables 1 and 2).

Of particular interest in interpreting these ^{18}O exchange studies is the finding that a carboxyl group on pepsin will incorporate two ^{18}O atoms from $H_2{}^{18}O$ without the addition of a virtual substrate (eq. 21) [170]. This

$$\text{Pepsin-COOH} + H_2{}^{18}O \xrightarrow[20°]{pH\,4.0} \text{Pepsin-C}^{18}O^{18}OH + H_2O \qquad (21)$$

exchange reaction does not occur if a diazo inhibited pepsin is used [171] suggesting that the exchangeable group is at the active site of pepsin. In the presence of Ac-Phe, an increase in incorporation of ^{18}O atoms, from 2 to 4, into the enzyme was observed. The evidence suggests that this incorporation might correspond to the complete exchange of four oxygen atoms from two carboxyl groups rather than partial exchange of several carboxyl groups. The most interesting result of this work is the finding that the rate constant for introduction of ^{18}O into pepsin is similar to that for exchange of Ac-Phe with $H_2{}^{18}O$.

Finally, Silver et al. [152] have shown that at pH > 4 the anionic form ($RCOO^\ominus$) of the virtual substrate and the half dissociated form of the enzyme $\left(E{\Large\langle}\begin{array}{l} COOH \\ COO^\ominus \end{array} \right)$ are the kinetically important species for the exchange reaction. A much slower exchange rate at lower pH's requires that an alternative path be available for reaction of $E{\Large\langle}\begin{array}{l} COO^\ominus \\ COOH \end{array}$ with the neutral form of the virtual substrate (RCOOH). It has been established that simple ester or amide derivatives of acyl amino acids do not participate in ^{18}O exchange reactions [152] nor act as acceptors in the transpeptidation reaction. This result can be accommodated by postulating that the carboxylate anion of the acyl amino acid ($RCOO^\ominus$) is the only species that pepsin will accept for the exchange and transpeptidation reaction. Silver et al. [152] suggest a mechanism to account for this ^{18}O exchange reaction as described by equations 22, 23, and 24. Equations 22 and 23 pertain to the ^{18}O exchange at high pH, whereas equations 23 and 24 account for the slower exchange at low pH. The rate-limiting step is either equation 22 or 24, followed in each case by a rapid step 23.

These mechanistic schemes for ^{18}O exchange are based on the premise that nucleophilic attack on a carboxylate anion is an unfavorable reaction. In simple organic systems, carboxylate salts undergo exchange only under rather drastic experimental conditions [172]. However, a recent report by Hegarty and Bruice [173] claims that N-(O-carboxyphenyl)urea cyclizes to 2,4-dihydroxyquinazoline at high pH ($>$ pH 11) by anionic attack upon a carboxylate anion. Although unlikely, a mechanism for O^{18} exchange by pepsin involving a nulceophilic attack on a carboxylate group can not be rejected at this time.

Silver et al. [152] list several advantages that the mechanism described by these equations possesses. Four that we find particularly interesting for ^{18}O exchange reactions are:

1. The spontaneous exchange of pepsin could arise by the carboxylate anion of pepsin replacing Ac-Phe-$COO^\ominus$ in equation 22.

2. For either slow step (eq. 22 or 24) the participating exchange species is the carboxylate anion (i.e., Ac-Phe-$COO^\ominus$).

3. These equations account for the fact that efforts to trap a pepsin-anhydride intermediate with methanol or hydroxylamine have failed. They predict that only the carboxyl group on the enzyme would be susceptible to esterification by methanol. If a methyl ester did form ($^\ominus O_2C$-E-$COOCH_3$) it should be very susceptible to a hydrolysis catalyzed by the free carboxylate group and therefore most difficult to isolate. Bruice and Benkovic [174] have

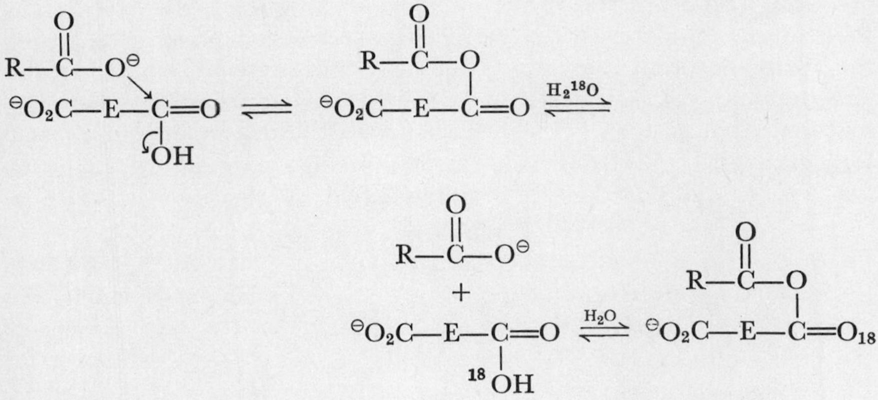

$$R-\underset{\overset{\parallel}{O}}{C}-O^{\ominus}$$

$$^{\ominus}O_2C-E-C=O$$

$$\underset{OH}{\overset{|}{}}$$

$\rightleftharpoons$

$$R-\underset{\overset{\parallel}{O}}{C}-O$$

$$^{\ominus}O_2C-E-C=O$$

$\overset{H_2^{18}O}{\rightleftharpoons}$

$$R-\underset{\overset{\parallel}{O}}{C}-O^{\ominus}$$
$$+$$
$$^{\ominus}O_2C-E-C=O$$
$$\underset{^{18}OH}{\overset{|}{}}$$

$\overset{H_2O}{\rightleftharpoons}$

$$R-\underset{\overset{\parallel}{O}}{C}-O$$
$$^{\ominus}O_2C-E-C=O_{18}$$

(22)

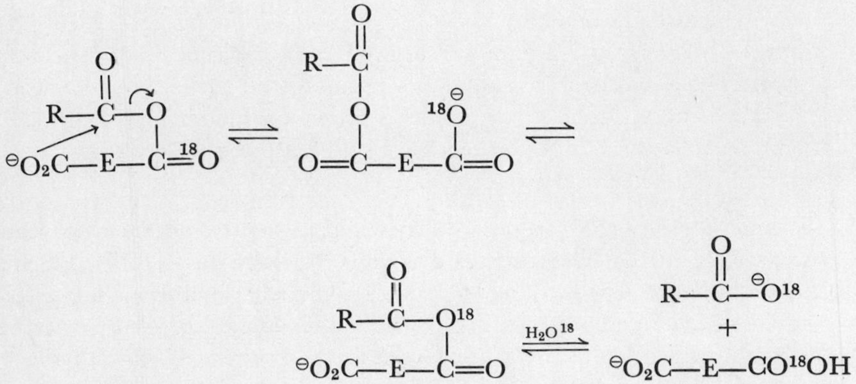

$$R-\underset{\overset{\parallel}{O}}{C}-O$$
$$^{\ominus}O_2C-E-C\overset{18}{=}O$$

$\rightleftharpoons$

$$R-\underset{\overset{\parallel}{O}}{C}$$
$$O \qquad {}^{18}O^{\ominus}$$
$$O=C-E-C=O$$

$\rightleftharpoons$

$$R-\underset{\overset{\parallel}{O}}{C}-O^{18}$$
$$^{\ominus}O_2C-E-C=O$$

$\overset{H_2O^{18}}{\rightleftharpoons}$

$$R-\underset{\overset{\parallel}{O}}{C}-{}^{\ominus}O^{18}$$
$$+$$
$$^{\ominus}O_2C-E-CO^{18}OH$$

(23)

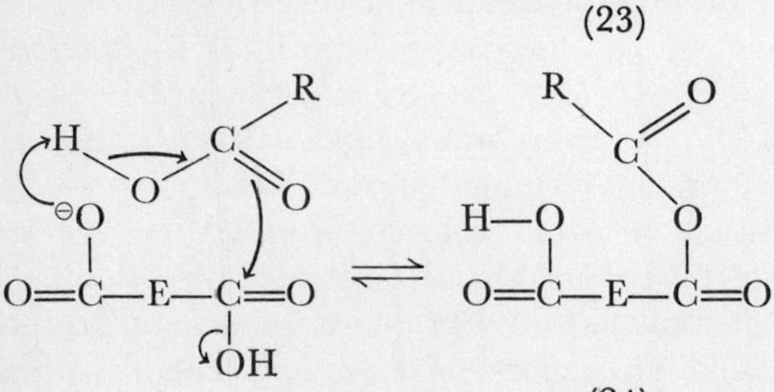

$$\overset{H}{}\quad \overset{C-R}{}$$
$$O \quad O$$
$$^{\ominus}O$$
$$O=C-E-C=O$$
$$\underset{OH}{\overset{|}{}}$$

$\rightleftharpoons$

$$R \quad O$$
$$C$$
$$H-O \qquad O$$
$$O=C-E-C=O$$

(24)

223

summarized the intramolecular carboxyl group catalysis reactions that could be models for this hydrolysis reaction.

4. The acyl-pepsin anhydride intermediate that is proposed as part of the pathway for ^{18}O exchange has likewise been suggested as being part of the hydrolysis mechanism (cf. eqs. 9 and 15). The rate of formation of this anhydride for ^{18}O exchange should be approximated by an intramolecular reaction. Higuchi et al. [175] showed that anhydrides do form in aqueous solution and their formation is the rate-limiting step in the following reaction (eq. 25). In addition the rate of cyclic anhydride formation in water in-

$$
\begin{array}{ccc}
& & O \\
& & \parallel \\
CH_2\text{—}COOH & CH_2\text{—}C & \\
| & \xrightarrow{\text{slow}} \quad | \quad \diagdown O \xrightarrow{\text{Aniline}} \\
CH_2\text{—}COOH & CH_2\text{—}C \diagup \\
& & \diagdown \\
& & O
\end{array}
\qquad
\begin{array}{c}
O \\
\parallel \\
CH_2\text{—}C\text{—}NH\phi \\
| \\
CH_2\text{—}COOH
\end{array}
\qquad (25)
$$

creased with the degree of methyl substitution in the following series [176]; 2, 3-dl-dimethylsuccinic > 2,2-dimethylsuccinic > methylsuccinic > succinic acid. Apparently, backside steric factors promote the cyclization reaction. Because the reaction rate is considerably more favorable for intramolecular reactions, they have been quite successfully used as models to approximate the reaction in the active site of enzymes.

May and Kaiser [151] proposed a mechanism for the pepsin catalyzed hydrolysis of sulfite esters consistent with the ^{18}O exchange [135] and pH dependency studies. As a working hypothesis, a mixed anhydride is suggested as a covalent intermediate (eq. 26). Since sulfite esters are quite reactive toward carboxylic acids [177] the proposed formation of a mixed anhydride is reasonable. Unfortunately, the mechanism for ^{18}O exchange shown in equation 22 is not easily reconciled with the mechanism for the sulfite ester hydrolysis (eq. 26). The role of carboxyl and carboxylate groups have been reversed in these reactions. If ^{18}O is incorporated only into the carboxyl group (eqs. 22 and 23) then equation 26 cannot account for the previously discussed ^{18}O incorporation into the bisulfite molecule (eq. 9).

Higuchi, McRae, and Shah [178] found that the rate of anhydride formation in aqueous solutions of succinic acid is significantly increased by the presence of sulfite species. Although the exact type of participation by sulfite in this reaction is unclear, it is interesting that sulfite esters are the most active substrates yet used with pepsin. The possibility of sulfite species acting as catalysts for pepsin hydrolyses of sulfite esters and peptides surely deserves further investigation. Such studies might provide additional

support for the anhydride intermediate as an integral part of the mechanism
of pepsin. The proposed anhydride-enzyme intermediates are by no means
unique to pepsin. There have been several recent reports [146, 179, 180]
that reactions of carboxypeptidase A proceed via an enzyme-substrate
anhydride intermediate (cf. section on D_2O effects).

5 MECHANISMS OF PEPSIN ACTION

To this point we have summarized the experimental observations that
must form the foundation for any proposed mechanism of pepsin action. As
pointed out repeatedly, many of the observations are difficult to interpret
from the mechanistic point of view, because they have been obtained under

(26)

a large variety of experimental conditions with enzymes of varying degrees of purity. For this reason, only a tentative mechanism for the action of pepsin is presented.

In formulating a mechanism of pepsin action, it would be almost impossible to account for every reported study. Therefore, the following discussion summarizes the aspects of pepsin action that we believe are on the firmest foundation.

1. At least two carboxyl groups are involved in pepsin catalyzed hydrolyses, and they are participating in the slow step.

a. The bell-shaped pH dependency curves suggest that the active form of the enzyme contains, in its active site, two carboxyl groups with pK_a's of about 1 and 4.7. These two carboxyls interact each other (possibly through hydrogen bonding) as reflected by their different pK_a values.

b. Negatively charged substrates and inhibitors bind poorly to pepsin suggesting a net negative charge or the negative end of an electrophile at the active site.

c. One aspartic acid side chain carboxyl group is inhibited by diazo compounds; another yet undetermined carboxyl group reacts specifically with epoxides.

d. The Michaelis or inhibitor constants for neutral substrates are generally pH independent. This suggests that the conformation of pepsin does not grossly change over the pH range of catalytic activity.

2. Pepsin substrate covalently bound intermediate(s) lie along the pathway of pepsin catalyzed hydrolyses.

a. The ordered release of products suggests a multistep catalytic reaction.

b. The observation of a D_2O solvent effect for one substrate and none for two other substrates suggests that there are at least two possible rate-limiting steps.

c. An amino-pepsin covalently bound intermediate is consistent with the transpeptidation reaction.

d. Although not on as firm a foundation, an acyl-pepsin intermediate is suggested by the ^{18}O exchange reaction. The anionic form of the virtual substrate (i.e., Ac-Phe-COO$^{\ominus}$) seems to be the important species in this exchange reaction.

Several theories have been proposed recently regarding the mechanism of pepsin action. However, none has received general support mainly because of the ambiguities and contradictions in the experimentation. Bender and Kézdy [181] began the mechanistic interpretation with the suggestion that two carboxylic acids on the enzyme participate in a four-center exchange reaction with the substrate, leading to an acyl-enzyme and an amino-enzyme intermediate. Bruice and Benkovic [182] suggested a similar concerted scheme. The essence of both proposals is incorporated in the mechanism [23] shown in equation 27. The following comments covering this mechanism are appropriate:

1. In the k_2 step the carboxylate acts as a nucleophile toward the carbonyl group of the amide bond and the k_2' step involves the concerted nucleophilic attack by the "liberated amine anion" on the carboxyl group of the enzyme. The k_2' step is similar to the reaction leading to the formation of N-methylphthalimide during the hydrolysis of N-methylphthalamic acid [183], which serves as an acceptable model for this first step in the mechanism. Another possible model for these steps (k_2 and k_2') is the postulated mechanism for the hydrolysis of actimycin A [184].

2. The k_3 step is the rapid hydrolysis of an anhydride and the k_4 step is the slower hydrolysis of an amide bond. A suitable model for the k_4 step would be the intramolecular hydrolysis of phthalamic acid [185] or succinanilic acid [186]. The hydrolysis of both of these compounds depends on an undissociated carboxyl group with a pK_a of 3 to 4. The intramolecular hydrolysis of phthalamic acid and succinanilic acid are 10^5 and 10^3 times faster than the hydrolysis of the unsubstituted benzamide or acetanilide, respectively.

3. This mechanism is consistent with equations 22, 23, and 24 for the pepsin catalyzed ^{18}O exchange with virtual substrates [152], since the protonated *carboxyl group* in pepsin is susceptible to *nucleophilic* attack by either the anion of the virtual substrate (i.e., Ac-Phe-COO$^\ominus$) or the "amide anion" of the substrate. If no peptide substrate is present, equation 22 accounts for the ^{18}O exchange, whereas steps k_3, k_3', k_{-3} and k_{-3}' account for this exchange if the carboxyl group is tied up in an amino-enzyme bond.

4. This mechanism also accounts for the observation [168] that at pH 4.4 the presence of a substrate (Ac-Phe-Tyr) increases the rate of incorporation of ^{18}O from water into (Ac-Phe-COO$^\ominus$). If k_{-3} or k_3' is of the same order of magnitude as k_4, then multiple exchanges between $H_2^{18}O$ and Ac-Phe-COO$^\ominus$ can occur before the enzyme intermediate reverts to final products.

5. This mechanism is consistent with the lack of success in trapping the anhydride intermediate. Equation 22 also requires that methanol can esterify only the carboxyl group on pepsin (cf. discussion in section on ^{18}O exchange).

6. If the two carboxyl groups in pepsin act together, as for example in a hydrogen bonded pair, there is no need in structure A to ionize the carboxyl group after hydrolysis of the anhydride. Once the other carboxyl group is engaged in bonding, the pK_a of the free carboxyl group need not be as low as it was in the free enzyme. For example, the 2 pK_a's in phthalic acid are 2.95 and 5.41, but in phthalamic acid the one pK_a is about 3.8.

7. If the first step (k_2) is rate limiting, then the absence of a D_2O solvent effect is expected because this step does not involve a proton transfer.

(27)

Evidence has been presented that suggests that the k_2 step is rate-determining for most of the pepsin substrates. For the cationic substrates, subsequent catalytic steps could contribute to the kinetically measured slow reaction. This would explain their more complex pH-rate profiles and the observation of a D_2O solvent isotope effect.

8. This mechanism is consistent with the requirement that the half dissociated form of pepsin ($^{\ominus}$OOC-E-COOH) and the anionic form (RCOO$^{\ominus}$) of the virtual substrate participate in the ^{18}O exchange reaction.

Knowles and co-workers have presented an alternate mechanism to account for hydrolysis, transpeptidation, and ^{18}O exchange [92a, 144, 187] as shown in equations 28 and 29. This mechanism is also closely related to the one suggested by Fruton and co-workers [102].

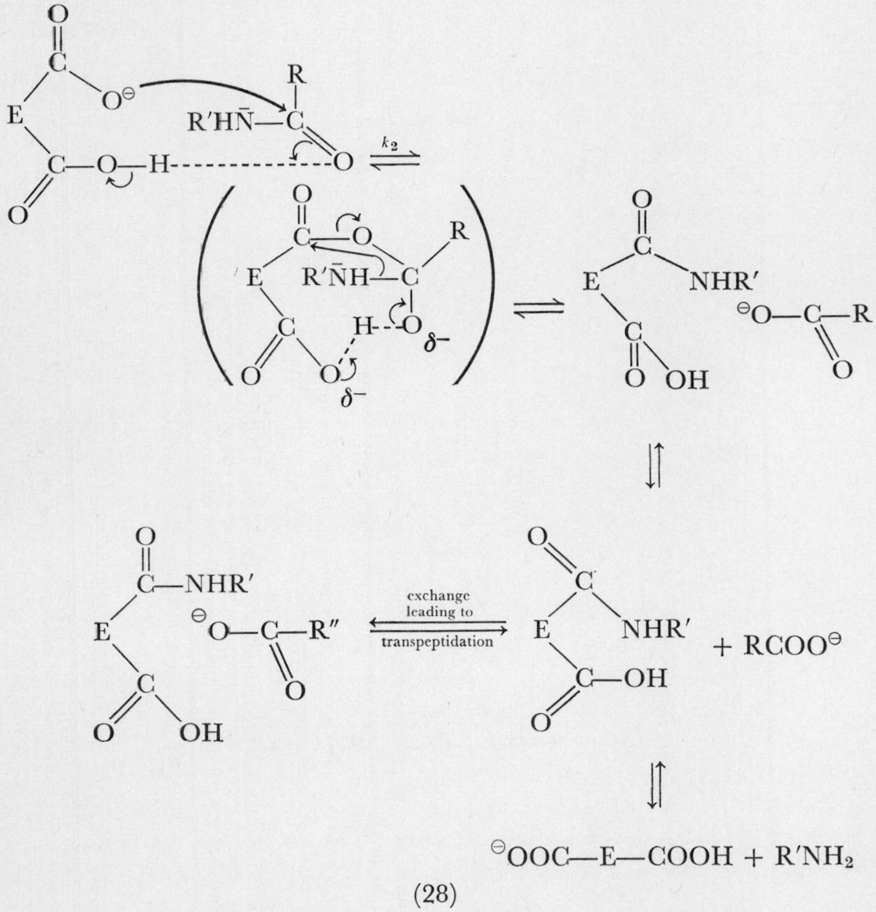

(28)

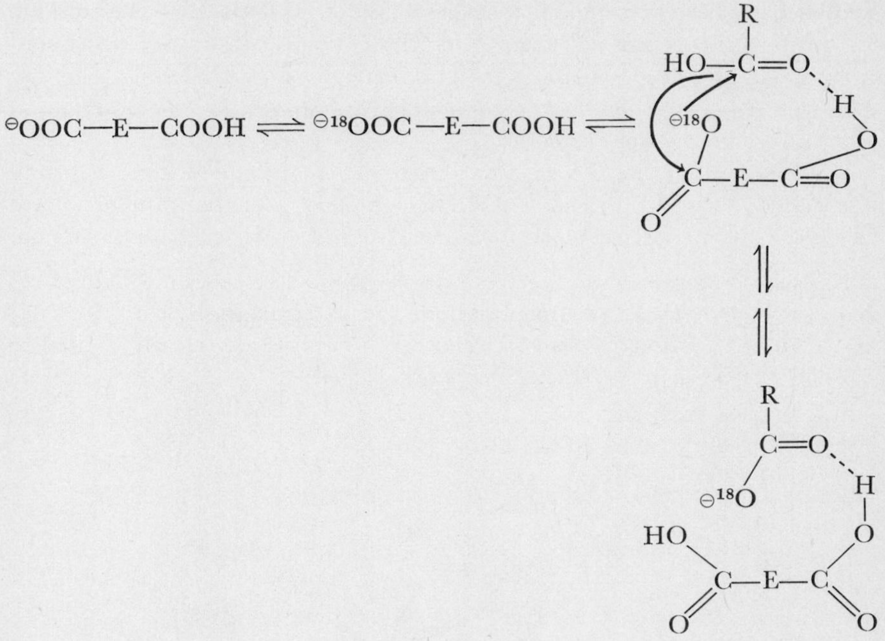

$$\Theta OOC-E-COOH \rightleftharpoons {}^{\Theta 18}OOC-E-COOH \rightleftharpoons \left(\begin{array}{c} \text{R} \\ | \\ HO-C=O \cdots H \\ {}^{\Theta 18}O \qquad O \\ C-E-C=O \\ O \end{array} \right.$$

$$\begin{array}{c} \text{R} \\ | \\ C=O \cdots H \\ {}^{\Theta 18}O \qquad O \\ HO \qquad | \\ C-E-C \\ O \qquad O \end{array}$$

(29)

The following points should be made concerning this mechanism:

1. An acyl pepsin anhydride intermediate is not formed during hydrolysis.

2. The first step k_2 is rate limiting because it involves a proton transfer; a D_2O solvent effect should accompany the hydrolytic reaction.

3. The prior release of the anionic acyl moiety occurs as demanded by their inhibition studies.

4. However, the ^{18}O exchange reaction (eq. 29) utilizes the protonated form of the acyl amino acid (RCOOH) instead of the anionic form observed by ^{18}O exchange studies [152].

5. The hydrolysis proceeds by the formation of a four-membered ring transition state, which we feel is not well documented.

6. The mechanism is not reconcilable with the results of the ^{18}O exchange during transpeptidation.

7. Knowles has postulated that pepsin must possess a more important electrophilic component in its mechanism then found for the neutral proteinases (e.g., α-chymotrypsin). It is apparent from examining Table 11 that the k ester/k amide hydrolysis ratio is large for those reactions proceeding

TABLE 11 RELATIVE RATES OF CATALYSED HYDROLYSIS OF ESTERS AND AMIDES

Catalyst	Ester	Amide	k_{ester}/k_{amide}
$OH^\ominus$	methyl acetate	acetamide	3400^a
$H_3O^\oplus$	methyl acetate	acetamide	9^a
α-chymotrypsin	N-acetyl-L-tryptophan methyl ester	N-acetyl-L-tryptopahn amide	1000^b
Pepsin	N-benzyloxycarbonyl-L-histidyl-p-nitro-L-phenylalanyl-L-β-phenyllactic acid methyl ester	N-benzyloxycarbonyl-L-histidyl-p-nitro-phenylalanyl-L-phenylalanine methyl ester	2^c
Carboxypeptidase A	N-benzoylglycyl-DL-β-phenyllactic acid	N-benzoylglycyl-L-phenylalanine	4^d

Source: This table was taken directly from reference 144.
^{a} Tables of Chemical Kinetics.
^{b} B. Zerner, R. P. M. Bond, and M. L. Bender, *J. Am. Chem. Soc.*, **86,** 3674 (1964).
^{c} Reference 64.
^{d} R. C. Davies, J. F. Riordan, D. S. Auld, and B. L. Vallee, *Biochemistry*, **7,** 1090 (1968).

via a nucleophilic mechanism. In contrast, this rate ratio is less than 10 for acid catalyzed hydrolysis reactions. The k ester/k amide hydrolysis ratio of 2 for pepsin was a compelling factor contributing to Knowles' postulate of an electrophilic proton transfer in the rate-limiting step of his mechanism. Considering pepsin's low pH optimum, it is difficult to argue against such a proton transfer reaction.

The mechanism of pepsin-catalyzed reactions can be compared to that of carboxypeptidase, which could proceed by a related electrophilic mechanism. The mechanism proposed by Kaiser and Kaiser [146] for the carboxypeptidase A catalyzed hydrolysis of peptides and esters begins with a nucleophilic attack on the carbonyl carbon to eventually yield an anhydride intermediate. The Zn ion acts as an electrophile to polarize the carbon-oxygen double bond and thereby promote nucleophilic attack. One of the less satisfactory aspects of the mechanism by Clement and the subsequent mechanism by Silver, Stoddard, and Stein [152] is the leaving or entering of $OH^\ominus$. Possibly one of the several carboxyl groups that were implicated by the inhibition studies [108] could participate as a third electrophilic functional group to facilitate the entering or leaving of this hydroxyl group. This group then would play the same role in pepsin as the zinc ion does in carboxypeptidase. That is, it contributes an electrophilic component to the mechanism of reaction. If such a group does exist and if it is a carboxyl group, its state of ionization could be outside the pH region of pepsin's activity. The actual positioning of the catalytically important groups in the active site of pepsin must await a determination of its linear sequence and three-dimensional structure.

An alternative hypothesis postulating an acyl enzyme and an amino-enzyme intermediate is shown in equation 30. The important features of this mechanism are summarized as follows:

1. The first step of the hydrolysis is similar to equation 27 involving nucleophilic attack by carboxylate group on the enzyme with $OH^{\ominus}$ as the leaving group. Arguments just presented support the idea of a third, as yet undetected, electrophilic group in the active site, which would facilitate this reaction.

2. The reaction then proceeds by way of an unsatisfying four-center reaction leading to an acyl-amino-pepsin intermediate.

3. This mechanism is consistent with the oxygen exchange mechanism (eq. 22).

4. It is unable to account for the increased ^{18}O exchange reaction that accompanies the transpeptidation.

5. The acyl and amino parts of the substrate become covalently linked to carboxyl groups *opposite* to those in equation 27.

6. The mechanism would preclude the successful trapping of the anhydride intermediate.

Finally, an additional hypothesis that has been recently offered is illustrated by equation 31 [188]. This equation assumes that pepsin has a slightly

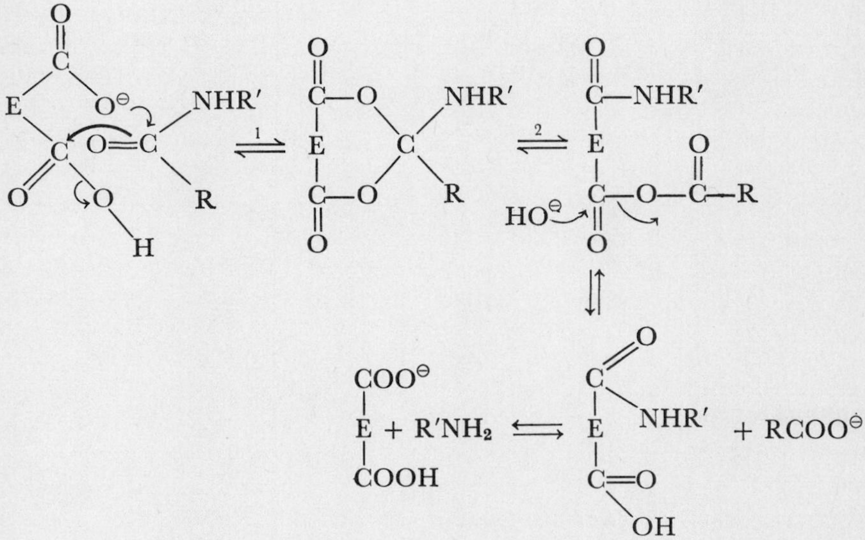

30

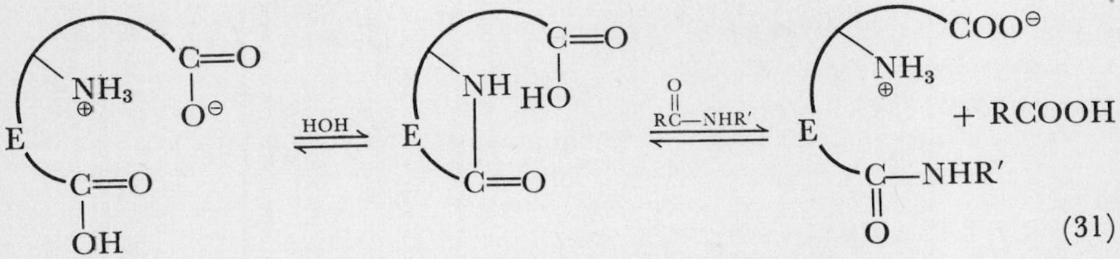

$$\xrightarrow{\text{HOH}}$$

$$\xrightarrow{\underset{\text{O}}{\overset{\parallel}{\text{RC—NHR}'}}}$$

$$\text{E} \qquad + \text{RCOOH} \tag{31}$$

distorted peptide linkage at its active site. A rapid protonation by the adjacent carboxyl group converts the carboxyl carbon atom into a strong electrophile that can react with either water or substrate. Because there is no experimental justification for the involvement of an amino group at the active site of pepsin, a more detailed consideration of this mechanism must await a completed linear sequence and/or three-dimensional structure.

6 CONCLUSION

The intent of this review is to stimulate additional interest in the proteolytic enzyme pepsin. If a consistent mechanism is to be found, we feel that more research incorporating *new approaches* is an absolute requirement. Throughout the chapter, areas demanding additional work have been emphasized, starting with the suggestion that only "pure" enzyme should be used. Apparently, the mechanism of pepsin action is sufficiently complex so that the uncertainties introduced by the use of impure pepsin must be eliminated. As was stated previously, every experimental area contributing to the mechanism of pepsin action requires further experimentation. Pepsin is a unique proteolytic enzyme operating in the low pH region. In the introduction we pointed out that pepsin is only one of a class of enzymes which exhibit their activity in this pH range. A more complete understanding of the mechanism of action of this class of enzymes would surely be an important contribution to our knowledge of enzymes in general. A specific enzyme in this class that has been particularly ignored is rennin. Preliminary studies on rennin [189] have shown a surprising similarity with pepsin. It is our belief that a comparison of pepsin with rennin will offer as many mechanistic advances, as has a similar comparison of chymotrypsin with trypsin.

REFERENCES

1. F. A. Bovey and S. S. Yanari, "The Enzymes," **4**, 63–92 (1960). An excellent review of pepsin prior to 1960.
2. J. H. Northrop, *J. Gen. Physiol.*, **13**, 739 (1930).
3. J. S. Fruton in *The Enzymes*, 3 *ed.* Vol. III, p. 119, Academic Press (1971).

4. D. J. Etherington and W. H. Taylor, *Bichem. Biophys. Acta*, **236,** 92 (1971).
5. D. J. Etherington and W. H. Taylor, *Biochem. J.*, **113,** 663 (1969).
6. M. D. Turner, J. C. Mangla, I. M. Samloff, L. L. Miller, and H. L. Segal, *Biochem. J.*, **116,** 397 (1970).
7. P. A. Meitner and B. Kassell, *Biochem. J.*, **121,** 249 (1971).
8. H. M. Lang and B. Kassell, *Biochemistry*, **10,** 2296 (1971).
9. T. P. Levchuk and V. N. Orekhovich, *Biokhimiva*, **28,** 1004 (1963).
10. S. T. Donta and H. VanVunakis, *Biochemistry*, **9,** 2791, 2798 (1970).
11. T. G. Merrett, E. Bar-Eli, and H. VanVunakis, *Biochemistry*, **8,** 3696 (1969).
12. W. Y. Huang and J. Tang, *J. Biol. Chem.*, **244,** 1085 (1969).
13. D. Tsuru, A. Hattori, H. Tsuji, and J. Fukumoto, *J. of Biochem.* **67,** 415 (1970).
14. J. Sodek and T. Hofmann, *Can. J. Biochem.*, **48,** 425 (1970).
15. B. Foltmann, *Compt. Rend. Trav. Lab. Carlsberg.*, **35,** 143 (1966).
16. B. Foltmann and B. S. Hartley, *Biochem. J.*, **104,** 1064 (1967).
17. B. Foltmann, *Phil. Trans. Roy. Soc. Lond.* **B257,** 147 (1970).
18. V. M. Stepanov, L. S. Lobareva, and N. I. Mal'tsev, *Biochim. Biophys. Acta*, **151,** 721 (1968).
19. R. D. Hill, *Biochim. Biophys. Res. Comm.*, **33,** 659 (1968).
20. M. K. Larson and J. R. Whitaker, *J. Dairy Sci.*, **53,** 262 (1970).
21. M. Bustin and A. Conway-Jacobs, *J. Biol. Chem.*, **246,** 615 (1971).
22. T. G. Rajagopalan, S. Moore, and W. H. Stein, *J. Biol. Chem.*, **241,** 4940 (1966).
23. G. E. Clement, S. L. Snyder, H. Price, and R. Cartmell, *J. Am. Chem. Soc.*, **90,** 5603 (1968).
24. R. Trujillo and M. Schlamowitz, *Anal. Biochem.*, **31,** 149 (1969).
25. V. M. Stepanov, E. A. Timokhina, and A. M. Zyakun, *Biochem. Biophys. Res. Comm.*, **37,** 470 (1969).
26. D. J. Etherington and W. H. Taylor, *Nature*, **216,** 279 (1967).
27. G. E. Clement, J. Rooney, D. Zahkeim and J. Eastman, *J. Am. Chem. Soc.* **92,** 186 (1970).
28. A. P. Ryle, *Biochem. J.* **96,** 6 (1965).
29. A. P. Ryle, *Biochem. J.*, **98,** 485 (1966).
30. A. P. Ryle and M. P. Hamilton, *Biochem. J.*, **101,** 176 (1966).
31. D. Lee and A. P. Ryle, *Biochem. J.*, **104,** 735, 742 (1967).
32. D. Lee, *Can. J. Biochem.*, **45,** 1002 (1967).
33. E. B. Ong and G. E. Perlmann, *J. Biol. Chem.*, **243,** 6104 (1968).
34. P. V. Koehn and G. E. Perlmann, *J. Biol. Chem.*, **243,** 6099 (1968).
35. J. Tang and B. S. Hartley, *Biochem. J.*, **118,** 611 (1970).
36. E. A. Vakhitova, M. M. Amirkhanyan, and V. M. Stepanov, *Biokhimiya*, **35,** 1164 (1970).
37. J. Tang, *Biochem. Biophys. Res. Comm.*, **41,** 697 (1970).
38. V. A. Trufanov, V. Kostka, B. Keil, and F. Sorm, *Eur. J. Biochem.*, **7,** 544 (1969).
39. V. M. Stepanov, M. M. Amirkhanyan, B. G. Belen'kii, R. A. Valyulis, E. A. Vakhitova, I. B. Pugacheva, and L. G. Senyutenkova, *Biokhimiya*, **35,** 283 (1970).
40. V. Kostka, L. Moravek, and F. Sorm, *Eur. J. Biochem.*, **13,** 447 (1970).
41. R. N. Perham and G. M. T. Jones, *Eur. J. Biochem.*, **2,** 84 (1967).
42. T. A. A. Dopheide, S. Moore, and W. H. Stein, *J. Biol. Chem.*, **242,** 1833 (1967).
43. V. M. Stepanov, V. I. Ostoslavskaya, V. F. Krivtzov, G. L. Muratova, and E. D. Levin, *Biochem. Biophys. Acta*, **140,** 182 (1967).
44. T. A. A. Dopheide and W. M. Jones, *J. Biol. Chem.*, **243,** 3906 (1968).
45. V. Kostka, I. Moravek, I. Kluh, and B. Keil, *Biochem. Biophys. Acta*, **175,** 459 (1969).

46. O. O. Blumenfeld and G. E. Perlmann, *J. Gen. Physiol.*, **42,** 553 (1959).
47. K. T. Fry, O. K. Kim, J. Spona, and G. A. Hamilton, *Biochemistry*, **9,** 4624 (1970).
48. J. Kay and A. P. Ryle, *Biochem. J.*, **123,** 75 (1971).
49. L. S. Shkarenkova and L. M. Ginodman, *Biokhimiya*, **33,** 1150 (1968).
50. M. L. Bender, M. L. Begue'-Canton, R. L. Blakely, L. J. Brubacher, J. Feder, C. R. Gunter, F. J. Kézdy, J. V. Killheffer, T. H. Marshall, C. G. Miller, R. W. Roeske, and J. K. Stoops, *J. Am. Chem. Soc.*, **88,** 5890 (1966).
51. G. R. Schonbaum, B. Zerner, and M. L. Bender, *J. Biol. Chem,* **236,** 2930 (1961).
52. J. P. Greenstein and M. Winitz, *Chemistry of Amino Acids*, Vol. 2, Wiley, New York, 1961, p.
53. G. Perlmann, *J. Biol. Chem.*, **241,** 153 (1966).
54. M. L. Anson, *J. Gen. Physiol.* **22,** 79 (1938).
55. M. D. Turner, J. L. Tuxill, L. L. Miller, and H. L. Segal, *Anal. Biochem.*, **16,** 487 (1966).
56. E. L. Gerring and E. A. Allen, *Clin. Chim. Acta*, **24,** 437 (1969).
57. L. E. Baker, *J. Biol. Chem.*, **193,** 809 (1951).
58. W. T. Jackson, M. Schlamowitz, and A. Shaw, *Biochemistry*, **4,** 1537 (1965).
59. J. Lenard, S. L. Johnson, R. W. Hyman, and G. P. Hess, *Anal. Biochem.*, **11,** 30 (1965).
60. T. R. Hollands and J. S. Fruton, *Biochemistry*, **7,** 2045 (1968).
61. A. J. Cornish-Bowden and J. R. Knowles, *Biochem. J.*, **96,** 71P (1965).
62. M. S. Silver, J. L. Denburg, and J. J. Steffens, *J. Am. Chem. Soc.*, **87,** 886 (1965).
63. G. W. Schwert and Y. Takenaka, *Biochem. Biophys. Acta.*, **16,** 570 (1955).
64. K. Inouye and J. S. Fruton, *Biochemistry*, **6,** 1765 (1967).
65. D. P. L. Sachs, E. Jellum, and B. Halpern, *Biochem. Biophys. Acta.*, **198,** 88 (1970).
66. T. P. Stein, T. W. Reid, and D. Fahrney, *Anal. Biochem.*, **41,** 360 (1971).
67. M. Bergmann and J. S. Fruton, *Advan. Enzymol.* **1,** 63 (1941).
68. E. J. Casey and K. J. Laidler, *J. Am. Chem. Soc.*, **72,** 2159 (1950).
69. M. Bergmann, *Advan. Enzymol.*, **2,** 49 (1942).
70. L. E. Baker, *J. Biol. Chem.*, **211,** 701 (1954).
71. R. L. Hill, *Advan. Protein Chem.*, **20,** 37 (1965).
72. W. T. Jackson, M. Schlamowitz, and A. Shaw, *Biochemistry*, **5,** 4105 (1966).
73. M. Schlamowitz and R. Trujillo, *Biochem. Biophys. Res. Comm.*, **33,** 156 (1968).
74. E. Zeffren and E. T. Kaiser, *J. Am. Chem. Soc.*, **89,** 4204 (1967).
75. J. L. Denburg, R. Nelson, and M. S. Silver, *J. Am. Chem. Soc.*, **90,** 479 (1968).
76. K. Inouye, I. M. Voynick, G. R. Delpierre, and J. S. Fruton, *Biochemistry*, **5,** 2473 (1966).
77. J. S. Fruton in "Structure-Function Relationships of Proteolytic Enzymes," Academia Press, N.Y. (1970) p. 222.
78. J. S. Fruton, *Advan. Enzymol.* **33,** 401 (1970).
79. T. R. Hollands, I. M. Voynick, and J. S. Fruton, *Biochemistry*, **8,** 575 (1969).
80. G. E. Trout and J. S. Fruton, *Biochemistry*, **8,** 4183 (1969).
81. G. E. Hein and C. Niemann, *J. Am. Chem. Soc.*, **84,** 4495 (1962).
82. K. Medzihradszyk, I. M. Voynick, H. Medzihradszky-Schweiger, and J. S. Fruton, *Biochemistry*, **9,** 1154 (1970).
83. G. P. Sachdev and J. S. Fruton, *Biochemistry*, **8,** 4231 (1969).
84. D. E. Koshland and K. E. Neet, *Ann. Rev. Biochem.*, **37,** 359 (1968).
85. W. P. Jencks in "Current Aspects of Biochemical Energetics," N. O. Kaplan and E. P. Kennedy, Eds., Academic Press, New York, 1966, p. 273.
86. G. E. Clement and S. L. Snyder, *J. Am. Chem. Soc.*, **88,** 5338 (1966).
87. J. R. Knowles, H. Sharp, and P. Greenwell, *Biochem. J.*, **113,** 343 (1969).
88. R E. Humphreys and J. S. Fruton, *Proc. Natl. Acad. Sci.*, U.S. **59,** 519 (1968).

89. M. S. Silver, *J. Am. Chem. Soc.*, **87**, 1627 (1965).
90. A. J. Cornish-Bowden, P. Greenwell, and J. R. Knowles, *Biochem. J.*, **113**, 369 (1969).
91. P. Greenwell, J. R. Knowles, and H. Sharp, *Biochem. J.*, **113**, 363 (1969).
92. T. M. Kitson and J. R. Knowles, *Biochem. J.*, **122**, 241 (1971); (a) *ibid.*, **122**, 249 (1971).
93. M. Schlamowitz, A. Shaw, and W. T. Jackson, *J. Biol. Chem.*, **243**, 2821 (1968).
94. W. T. Jackson, M. Schlamowitz, A. Shaw, and R. Trujillo, *Arch. Biochem. Biophys.*, **131**, 374 (1969).
95. J. Tang, *Nature*, **199**, 1094 (1963).
96. J. Tang, *J. Biol. Chem.*, **240**, 3810 (1965).
97. K. Inouye and J. S. Fruton, *Biochemistry*, **7**, 1611 (1968).
98. R. Y. Hsu, W. W. Cleland and L. Anderson, *Biochemistry*, **5**, 799 (1966).
99. J. R. Knowles, *J. Theor. Biol.*, **9**, 213 (1965).
100. J. H. Northrop, *J. Gen. Physiol.*, **5**, 263 (1922).
101. R. M. Herriott, *J. Cell. Comp. Physiol.*, **47**, Suppl. 1, 239 (1956).
102. C. R. Delpierre and J. S. Fruton, *Proc. Natl. Acad. Sci.*, **54**, 1161 (1965).
103. G. R. Delpierre and J. S. Fruton, *Proc. Natl. Acad. Sci., U.S.*, **56**, 1817 (1966).
104. G. A. Hamilton, J. Spona, and L. D. Crowell, *Biochem. Biophys. Res. Comm.* **26**, 193 (1967).
105. K. T. Fry, Oh-Kil-Kim, J. Spona, and G. A. Hamilton, *Biochem. Biophys. Res. Comm.*, **30**, 489 (1968).
106. V. M. Stepanov and T. I. Vaganova, *Biochem. Biophys. Res. Comm.* **31**, 825 (1968).
107. T. G. Rajagopalan, W. S. Stein, and S. Moore, *J. Biol. Chem.*, **241**, 4295 (1966).
108. W. H. Stein in "Structure Function Relationships of Proteolytic Enzymes," Academic Press, New York, 1970, p. 253.
109. R. L. Lundblad and W. H. Stein, *J. Biol. Chem.*, **244**, 154 (1969).
110. E. B. Ong and G. E. Perlmann, *Nature*, **215**, 1492 (1967).
111. R. S. Bayliss and J. R. Knowles, *Chem. Comm.*, 196 (1968).
112. R. S. Bayliss, J. R. Knowles, and G. B. Wybrandt, *Biochem. J.*, **113**, 377 (1969).
113. L. V. Kozlov, L. M. Ginodman, and V. N. Orekhovich, *Biokhimiya*, **32**, 1011 (1967).
114. B. F. Erlanger, S. M. Vratsanos, N. Wassermann, and A. G. Cooper, *J. Biol. Chem.*, **240**, PC 3447 (1965).
115. B. F. Erlanger, S. M. Vratsanos, N. Wassermann, and A. G. Cooper, *Biochem. Biophys. Res. Comm.*, **23**, 243 (1966).
116. E. Gross and J. L. Morell, *J. Biol. Chem.*, **241**, 3638 (1966).
117. B. F. Erlanger, S. M. Vratsanos, N. Wassermann, and A. G. Cooper, *Biochem. Biophys. Res. Comm.*, **28**, 203 (1967).
118. S. S. Husain, J. B. Ferguson, and J. S. Fruton, *Proc. Natl. Acad. Sci. U.S.*, **68**, 2765 (1971).
119. J. Tang, *J. Biol. Chem.*, **246**, 4510 (1971).
120. D. G. Hoare and D. E. Koshland, *J. Am. Chem. Soc.*, **88**, 2057 (1966).
121. B. M. Trost, *J. Am. Chem. Soc.*, **84**, 138 (1967).
122. E. Gross in "Structure Function Relationships of Proteolytic Enzymes," Academic Press (1970) p. 252. P. Desnuelle, H. Neurath and M. Ottesen, editors.
123. W. C. J. Ross, *J. Chem. Soc.*, 2257 (1960).
124. R. M. Herriott, *J. Gen. Physiol.*, **19**, 283 (1935).
125. L. A. Lokshina and V. N. Orekhovich, *Biokhimiya*, **31**, 143 (1966).
126. L. V. Kozlov, G. A. Kogan, and L. L. Zavada, *Biokhimiya*, **34**, 1257 (1969).
127. S. Rimon and G. E. Perlmann, *J. Biol. Chem.*, **243**, 3566 (1968).
128. L. A. Lokshina and V. N. Orekhovich, *Biokhimiya*, **29**, 346 (1964).
129. R. A. Badley and W. J. Teale, *Biochem. J.*, **116**, 341 (1970).

130. Y. Nakagawa and G. E. Perlmann, *Arch. Biochem. Biophys.*, **144**, 59 (1971).
131. L. A. Lokshina, V. N. Orekhovich, and V. A. Sklyankina, *Nature*, **204**, 580 (1964).
132. K. Inouye and J. S. Fruton, *J. Am. Chem. Soc.*, **89**, 187 (1967).
133. T. W. Reid and D. Fahrney, *J. Am. Chem. Soc.*, **89**, 3941 (1967).
134. T. W. Reid, T. P. Stein, and D. Fahrney, *J. Am. Chem. Soc.*, **89**, 7125 (1967).
135. T. P. Stein and D. Fahrney, *Chem. Comm.*, 555 (1968).
136. S. W. May and E. T. Kaiser, *J. Am. Chem. Soc.*, **91**, 6491 (1969).
137. C. D. Hubbard and T. P. Stein, *Biochem. Biophys. Res. Comm.*, **45**, 293 (1971).
138. S. W. May and E. T. Kaiser, *J. Am. Chem. Soc.*, **93**, 5567 (1971).
139. A. J. Cornish-Bowden and J. R. Knowles, *Biochem. J.*, **113**, 353 (1969).
140. L. Peller and R. A. Alberty, *J. Am. Chem. Soc.*, **81**, 5907 (1959).
141. B. Zerner and M. L. Bender, *J. Am. Chem. Soc.*, **83**, 2267 (1961).
142. M. L. Bender, G. E. Clement, F. J. Kézdy, and H. d'A. Heck, *J. Am. Chem. Soc.*, **86**, 3680 (1964).
143. T. R. Hollands and J. S. Fruton, *Proc. Natl. Acad. Sci.* U.S., **62**, 1116 (1969).
144. J. R. Knowles, *Phil. Trans. Roy. Soc. Lond.*, **B257**, 135 (1970).
145. B. L. Kaiser and E. T. Kaiser, *Proc. Natl. Acad. Sci.*, U.S., **64**, 36 (1969).
146. B. L. Kaiser and E. T. Kaiser, Accounts Chem. Res., **5**, 219 (1972).
147. H. Neumann, Y. Levin, A. Berger, and E. Katchalski, "Proceedings of the International Symposium on Enzyme Chemistry, Tokyo-Kyoto (1957) (K. Ichihara, ed.) p. 129, Academic Press, New York (1958); *Biochem. J.*, **73**, 33 (1959).
148. N. Sharon, V. Grisaro, and H. Neumann, *Arch. Biochem. Biophys.*, **97**, 219 (1962).
149. D. G. Doherty and F. Vaslow, *J. Am. Chem. Soc.*, **74**, 931 (1952).
150. V. Grisaro and N. Sharon, *Biochem. Biophys. Acta*, **84**, 152 (1964).
151. S. W. May and E. T. Kaiser, *Biochemistry*, **11**, 592 (1972).
152. M. S. Silver, M. Stoddard, and T. P. Stein, *J. Am. Chem. Soc.*, **92**, 2883 (1970).
153. M. L. Bender, G. E. Clement, C. R. Gunter, and F. J. Kézdy, *J. Am. Chem. Soc.*, **86**, 3697 (1964).
154. C. Lowe and A. Williams, *Biochem. J.*, **96**, 199 (1965).
155. M. Akhtar and J. M. Al-Janabi, *Chem. Comm.*, 1002 (1969).
156. M. Akhtar, *Chem. Comm.*, 361 (1970).
157. T. M. Kitson and J. R. Knowles, *Chem. Comm.*, 361 (1970).
158. C. Godin and C. Y. Yuan, *Chem. Comm.*, 84 (1970).
159. M. Akhtar and J. M. Al-Janabi, *Chem. Comm.*, 859 (1969).
160. J. S. Fruton, S. Fujii, and M. H. Knappenberger, *Proc. Natl. Acad Sci.* U.S., **47**, 759 (1961).
161. L. V. Kozlov, L. M. Ginodman, V. N. Orekhovich, and T. A. Valueva, *Biokhimiya*, **31**, 315 (1966).
162. M. S. Silver and M. Stoddard, *Biochemistry*, **11**, 191 (1972).
163. N. I. Mal'tsev, L. M. Ginodman, V. N. Orekhovich, T. A. Valueva, and L. M. Akimova *Biokhimiya*, **31**, 983 (1966).
164. H. Newmann and N. Sharon, *Biochem. Biophys. Acta*, **41**, 370 (1960).
165. M. L. Bender and F. J. Kézdy, *J. Amer. Chem. Soc.*, **86**, 3704 (1964).
166. G. E. Clement, A. Siegel, and R. Potter, *Can. J. Biochem.*, **49**, 477 (1971).
167. L. V. Kozlov, L. M. Ginodman, B. M. Zolotarev, and V. N. Orekhovich, *Dokl. Akad. Nauk SSSR*, **146**, 945 (1962).
168. L. V. Kozlov, L. M. Ginodman, and V. N. Orekhovich, *Dokl. Akad. Nauk SSSR*, **161**, 1455 (1965).
169. L. V. Kozlov, L. M. Ginodman, and V. N. Orekhovich, *Dokl. Akad. Nauk SSSR*, **172**, 1207 (1967).

170. L. S. Shkarenkova, L. M. Ginodman, L. V. Kozlov, and V. N. Orekhovich, *Biokhimiya*, **33,** 154 (1968).
171. L. M. Ginodman, T. A. Valueva, L. V. Kozlov, and L. S. Shkarenkova, *Biokhimiya*, **34,** 211 (1969).
172. D. Samuel and B. L. Silver, *Advan. Phys. Org. Chem.*, **3,** 168 (1965).
173. A. F. Hegarty and T. C. Bruice, *J. Am. Chem. Soc.*, **91,** 4924 (1969).
174. T. C. Bruice and S. Benkovic, "Bio-organic Mechanisms," Vol. I, W. A. Benjamin, Inc., New York, N.Y., 1966, pp. 173–186.
175. T. Higuchi, T. Niki, A. C. Shah, and A. K. Herd, *J. Am. Chem. Soc.*, **85,** 3655 (1963).
176. T. Higuchi, L. Eberson, and J. D. McRee, *J. Am. Chem. Soc.*, **89,** 3001 (1967).
177. B. Iselin, W. Rittel, P. Sieber, and R. Schwyzer, *Helv. Chim. Acta*, **40,** 373 (1957).
178. T. Higuchi, J. D. McRae, and A. C. Shah, *J. Am. Chem. Soc.*, **88,** 4015 (1966).
179. W. N. Lipscomb *et al.*, *Brookhaven Symp. Biol.*, **21,** 24 (1969).
180. B. L. Vallee and J. F. Riordan, *Brookhaven Symp. Biol.*, **21,** 91 (1969).
181. M. L. Bender and F. J. Kézdy, *Ann. Revs. Biochem.*, **34,** 49 (1965).
182. Reference 174, pp. 2–4.
183. J. Brown, S. C. K. Su, and J. A. Shafer, *J. Am. Chem. Soc.*, **88,** 4468 (1966).
184. A. J. Birch, D. W. Cameron, R. W. Richards, and Y. Harada, *Proc. Chem. Soc.*, 22 (1960). E. E. VanTamelen, *J. Am. Chem. Soc.*, **83,** 1639 (1961).
185. M. L. Bender, Y. Chow, and F. Chloupek, *J. Am. Chem. Soc.*, **80,** 5380 (1958).
186. T. Higachi, L. Eberson, and A. K. Herd, *J. Am. Chem. Soc.*, **88,** 3805 (1966).
187. J. R. Knowles, R. S. Bayliss, A. J. Cornish-Bowden, P. Greenwell, T. M. Kitson, H. C. Sharp, and G. B. Wybrandt, in "Structure Function Relationships of Proteolytic Enzymes," P. Desnuelle, H. Neurath, and M. Ottesen Eds., Academic Press, New York, 1970, p. 237.
188. J. H. Wang *ibid.*, p. 251.
189. G. E. Clement and G. S. Cashell, *Biochem. Biophys. Res. Comm.*, **47,** 328 (1972).

AUTHOR INDEX

Numbers in parentheses are reference numbers and show that an author's work is referred to although his name is not mentioned in the text. Numbers in *italics* indicate the pages on which the full references appear.

SUBJECT INDEX

CUMULATIVE INDEX, VOLUME I AND 2